# The Role of Horticulture
# in Human Well-Being
# and
# Social Development

# The Role of Horticulture in Human Well-Being and Social Development:

## A NATIONAL SYMPOSIUM

*19–21 April 1990—Arlington, Virginia*

Editor-in-chief

## Diane Relf

Associate Professor of Horticulture,
Virginia Polytechnic Institute and State University

**TIMBER PRESS**
*Portland, Oregon*

Bronze sculptures following page 112 are by Rhonda Roland Shearer; copyright of Rhonda Roland Shearer; courtesy of the Wildenstein Gallery, 19 East 64th Street, New York, NY 10021. For information on the sculptures contact Ms. Joyce Hartke at the Wildenstein Gallery, (212) 879-0500.

Reprinted 1994

ISBN 0-88192-209-9
Printed in Singapore

TIMBER PRESS, INC.
9999 S.W. Wilshire, Suite 124
Portland, Oregon 97225

Library of Congress Cataloging-in-Publication Data

The Role of horticulture in human well-being and social development :
    a national symposium, 19-21 April 1990, Arlington, Virginia / editor
    -in-chief, Diane Relf.
        p.    cm.
    Includes bibliographical references and index.
    ISBN 0-88192-209-9
    1. Horticulture--Congresses.   2. Horticulture--Social aspects-
    -Congresses.   3. Gardening--Therapeutic use--Congresses.   I. Relf,
    Diane.
    SB317.53.R65   1992
    635--dc20                                              91-19911
                                                              CIP

# *Contents*

———————

## SECTION III. PLANTS AND THE INDIVIDUAL

## SECTION IV. DEVELOPING A CONCEPTUAL FRAMEWORK

SECTION V. EXPLORING A SPECIFIC APPLICATION: HORTICULTURAL THERAPY

# Dedication

If it is appropriate to dedicate a symposium to an individual, then this one should certainly be dedicated to **Charles Lewis**. He has been bringing the people/plant interaction message to horticulture for over 20 years, and it was he who provided the inspiration for this initiative.

# *Foreword*

———————

The essence of this multidisciplinary symposium lies in recognizing, understanding, and appreciating the differences in the way that each of us perceives the world around us. Each person's view is unique, shaped by one's own personality and experiences. For example, a geranium can be source of many meanings. A taxonomist will see a member of the family *Geraniaceae*, and examine minute details of various parts of the plant to decide on its exact classification. Another group, plant breeders, interested in enhancing the plant's ornamental qualities through hybridization, will look for variation in color and form of leaf and flower, variation in growth habit; physiologists might want to create the most efficient conditions for growth through control of environmental conditions. A commercial grower is primarily concerned with how to grow the geranium and sell it for a profit. The geranium must represent sufficient value for the customer that he be willing to purchase it.

The kinds of meaning a geranium might convey depend, in part, on the setting in which it is ultimately grown. In a flourishing suburban garden, it might hold a different meaning than one grown in a window box in a low-income housing area, by a patient in a nursing home or hospital, in a school for the mentally disadvantaged, a physical rehabilitation center, or a prison. It is the same plant; but each person will endow the plant with his or her own interpretation and meaning. To paraphrase Gertrude Stein: A geranium is a geranium is a geranium.

I hope that in this symposium, each of us can, for a moment, abandon the limitations of our own viewpoint to appreciate the diverse meanings people find in plants. We are here today to build bridges, help each other understand and appreciate this diversity of view.

To fully appreciate the significance of its plants, the horticulture community needs to join with psychology and sociology, the sciences of people, to gain a view of the role played by our plants in the world of people. In doing so, they will restore to horticulture its human context. The word is derived from the Latin, *hortus*, garden, and *cultura*, to cultivate. This meaning was obscured by academia when it shifted to a scientific focus for horticulture.

Why might this human context be important for horticulture? If people/plant interactions did not occur, and people were not attracted to plants, and not willing to exchange money for them, could commercial horticulture exist? Therefore, this conference bears directly on the well-being of commercial ornamental horticulture.

The human context becomes more evident when we examine the content of horticulture. If asked what gardening and horticulture are all about, we might quickly respond—plants! But think about it: For whose benefit are horticultural books written, botanical societies, plant societies, and garden clubs established? Were Latin plant names invented so that plants would know who they were, or so that people could communicate with each other about specific plants? Are plants sought, discovered, researched, propagated, bred, bought and sold for the benefit of plants or to satisfy the interests of people? Plants do not need people; but indeed, people do need plants. Horticulture and gardening are essentially human activities; they serve as an umbrella to bring together all people who have an interest in plants.

Charles Lewis
Morton Arboretum, June 1990

# Introduction

This first-of-its-kind symposium was an important step toward what Dr. Jules Janick speaks of as the "new role of horticulture to re-establish the bond between plants and people." In studying these proceedings, it is important to focus on the ultimate goal of the symposium: to establish a research initiative on Human Issues in Horticulture and, with that initiative, a system to communicate the research to a comprehensive audience and to work toward the application of the research findings.

A research initiative on Human Issues in Horticulture is needed to document clearly through research the value to people of active and passive experiences with plants in order to implement actions that will allow the greatest number of people to garner these benefits. Expressed in a different way, the goal is for horticulture/gardening/ greening/people-plant interaction to be to human well-being in the 1990s what sports/physical fitness/health/exercise was in the 1980s.

The horticulture community, as we are discussing it here, consists of all professionals whose economic well-being is dependent upon horticulture and its related activities, including, but not limited to, staff of arboreta and botanic gardens; faculty and staff in university horticulture departments; trade, amateur, and professional associations; members of the horticulture industry; and other educators and practitioners such as community garden leaders, city horticulturists, horticultural therapists, and garden writers. It also includes amateurs who view their horticultural activities as an important contribution to society (such as Master Gardeners). Arboreta and botanic gardens have a role in human social development, as is discussed by Sue Brogdon of the Chicago Botanic Gardens. They certainly have a role in restoring some peace and tranquility to individuals as they escape from the daily rush into the oasis of these beautiful gardens. Visiting an arboretum or botanic garden is an important human cultural experience. As part of America's museum system, they can expand their cultural role by demonstrating that plants are integral to other human cultural experiences, from music and literature to poetry and fine arts—as we see in the sculptures of Rhonda Roland Shearer. Arboreta and botanic gardens are more than repositories for plants—Rene Dubois wrote that they were conceived as a way to restore the Garden of Eden by collecting all the plants of the world together again. If that is the case, what roles are arboreta and botanical gardens to play in the lives of the "children of Adam and Eve"?

Horticulture departments at universities have a mandate to teach, conduct research, and carry information on horticulture to the public. As early as 1970, D. Spiegel-Roy, President of the International Horticultural Congress, stated that "the aim of horticulture . . . a science with a human goal . . . must be to make human life healthier and more enjoyable, to provide man with beauty, color, and form." Only by going beyond questions of commercial crop production and the search for basic scientific knowledge on plant growth can we fulfill our mandate in today's society.

In 1972 [*HortScience* 7(6):544], Sylvan H. Wittwer, Professor of Horticulture at Michigan State University, in response to the tremendous growth in the number of students in horticulture across the country at that time, wrote, "Are we equal to the challenge? Is there a message? Will we pass up this opportunity? Horticulture for the millions should be our motto." He set forth a challenge: "Restructuring of investments in research and training is called for both at the graduate and undergraduate levels. Greater opportunities should be provided in urban and landscape horticulture; recreational, environmental, and therapeutic horticulture; in fruit, vegetable and flower gardening; in addition to meeting traditional needs."

In 1976 [*HortScience* 11(1):4–5], Charles Lewis, who was then the Coordinator of the American Horticultural Society's People/Plant Program, wrote, "Society has found horticulture. With the people/plant concept, horticulture can discover new and vital dimensions in society. The questions concerning people/plant interaction will be answered because the pressures of human needs demand answers. To what degree will horticulture participate in the search? Can we enlarge the area of our horticultural concerns to include inherent human benefits?"

In 1983 [*HortScience* 18(1):11–13], Harold Tukey, the Director of the Center for Urban Horticulture, University of Washington, wrote *with such accuracy*, "It is long past time when horticulturists should combine forces with the psychologist, the artist, and the landscape architect to quantify in scientific terms the effects that plants have on humans in addition to providing food and substance."

The 1990s promise to be the decade of the environment. One of the most effective ways to address environmental issues on an individual basis is through horticultural activities such as planting trees and gardens. The environmental concerns of the 1990s will translate into unique opportunities for the horticultural community having an impact on society and quality of life if we have the data to substantiate the benefits of proximity to plants that people feel they already "know." Data from research in human issues in horticulture will provide significant impact both in terms of cost/benefit analysis and visibility through media and research report channels.

It is important that we be ready to fund, conduct, communicate, and apply research to ensure growth of the horticultural community and the subsequent impact on the quality of human life.

So we from the horticulture community are here—however belatedly—no longer to issue challenges, but to take action; to learn from our colleagues in forestry, psychology, art, architecture, urban planning, sociology, and history the knowledge their research has provided, to report on our beginning research efforts, and to develop linkages for future research. In speaking to the symposium participants gathered for the reception at the United States Botanic Garden, Charles Hess, USDA Assistant Secretary for Science and Education, offered these words of support and guidance:

> As those of you here at this symposium clearly recognize, increased investment in agricultural research is a necessary investment in the future if agriculture is to continue to be competitive and to be environmentally sensitive.
>
> Through research, we offer America's horticulturists new ways to produce quality products efficiently and to expand their markets. We are moving into the future and through your efforts are broadening the entire vision of what horticulture can and should be.

It has become clear at most universities that we can no longer be divided by specialties. To meet today's challenges, we must "broaden our vision" and work cooperatively in interdisciplinary research teams. Some horticulturists are already reaching out by becoming members of teams that are concerned with human physical, psychological, and social well-being; teams that are seeking answers to ameliorate the stresses of modern life. Horticulture therapy is one area in which progress has been made, but other relevant areas have been left unexplored by horticulturists. The work place, the community, the retirement center, even resorts and theme parks are prime areas for the involvement of horticulturists in research. Other disciplines are already conducting research on people's response to vegetation, with researchers ready to interact with horticulturists in understanding the role of plants. A few of these disciplines are architecture, building engineering, gerontology, hotel management, social sciences, and humanities.

Another essential element of the horticultural community is strongly behind this initiative to understand human issues in horticulture—the horticulture industry. This symposium and resulting activities have been endorsed by nearly all of the major horticultural trade/professional associations. Funds have been supplied by horticulture foundations and industry leaders, including the Horticultural Research Institute, the American Floral Endowment, the Kenneth Post Foundation, the Associated Landscape Contractors of America, the Florida Nurserymen and Growers Association, and more than ten private firms. The Garden Writers Association of America has provided initial funding to ensure publication of the symposium proceedings. Larry Scovotto, Executive Vice-President of the American Association of Nurserymen, explains the importance of this initiative to the industry:

> The horticulture industry is in the enviable position of marketing a product that is good for people's mental and physical well-being. And the more that it is used, the better it is for the environment. Research which is focused on these issues can serve only to benefit both our industry and the users of horticultural products.
>
> People/plant or human issues research means a more effective utilization of the resources that are part of our cost of doing business. To the ultimate individual user of horticultural products, this research can mean an improvement in the quality of . . . life. And to our industry, this research will be invaluable in identifying additional suitable plants and in the development of effective management programs for plants in man-made settings.
>
> Quite clearly, a better understanding of peoples' response to plants will increase the use of plants; which is our bottom line. However, all research findings on people/plant interaction will not be positive. We in the horticulture industry must be ready to adjust our practices to optimize the positive benefits. We need to take a leadership role in encouraging and funding research in this area, in communicating the research findings to the public, and in establishing programs to ensure the application of these research findings.

Certainly participation by the horticulture industry is an essential element for the growth of horticulture into new and nontraditional areas. As research documents the benefits of plants to the quality of human life and educators and communicators carry the message to Dr. Wittwer's "millions," it is the role of commercial horticulturists to supply the crops and services that will make possible the "life-enhancing" impact of plants. Suppliers, growers, wholesalers, florists, nurserymen, landscapers, and other professionals whose economic well-being is dependent on people's involvement and satisfaction with plants will have a significant role in implementing the findings of research focused on horticulture and human well-being; and, obviously, they have much to gain from this.

In structuring this symposium, we have attempted to survey broadly the Human Issues in Horticulture to develop an overview of how plants affect people and to make clear to members of the horticulture community the diverse opportunities for research and acquisition of new knowledge.

Researchers from the fine arts, sociology, psychology, urban planning, forestry, environmental psychology, and history are among those who participated in this symposium, ready to conduct interdisciplinary work with members of the horticulture community. Let them help you identify your personal area of commitment to this research—to conducting it, cooperating on it, funding it, communicating it, or implementing it. The time for challenging others has passed. We are here to commit. We are here to act.

Diane Relf
Virginia Polytechnic Institute and State
University, June 1990

# SECTION I

# PLANTS AND HUMAN CULTURE

CHAPTER 1

# Horticulture and Human Culture

Jules Janick

Department of Horticulture, Purdue University, West Lafayette, Indiana
(James Troop Distinguished Professor in Horticulture)

## INTRODUCTION

Earth is a plant-oriented planet. The green plant is fundamental to all other life. Were humanity to perish tomorrow, vines would destroy our mighty temples and grass would soon grow in the main streets of the world. In contrast, the disappearance of plants would be accompanied by the disappearance of humankind along with every other animal.

The importance we attribute to any product, however, is related to the probability or actuality of a shortage rather than intrinsic value. Thus, those that are plentiful and readily available are often held in low esteem, even though our very existence may depend upon them. The oxygen we breathe, the nutrients we consume, the fuels we burn, many of the most important materials we use, are all related to plant life.

Plants have a wide spectrum of uses. The most obvious is for human sustenance. Plants supply all of our food, either directly or indirectly as feed for animal intermediaries. They are also utilized as a source of structural support, as a construction material, and as the raw material in the manufacture of fabrics and paper and such synthetics as plastics and rayon. We have come to depend upon many of the complex substances that plants produce—dyes, tannins, waxes, resins, flavorings, medicines, and drugs. Living plants, besides having a direct effect on the ecological position of humans, are used to control erosion by water and wind, to provide a setting for recreation and sport, as landscape materials, and to satisfy our desire for beautiful objects.

The story of humanity is largely a chronicle of our struggle for dominion over the environment. The efficiency of this control is thought of in terms of civilization, or culture. To a great extent, controlling the environment means controlling plant life. The failure to provide a high standard of living for the population as a whole in so-called underdeveloped countries is generally associated with an inefficient system of crop production and distribution. Similarly, the decline of developed societies can in many instances be associated with a disruption in the basic system of resource utilization. Crop production, the management of useful plants, is the very basis of our civilization.

The perceived relationship of plants and people has changed throughout human history and is still changing. Let us trace three cultural waves: pre-agricultural, agricultural, and industrial, and then step into the future.

## PRE-AGRICULTURAL

Human beings have been hunter-gatherers for 99% of the 2 million years our race has roamed the earth; only in the last 10,000 years have people been agriculturists. Jack R. Harlan, in his book *Crops and Man*, summarized the life of pre-agricultural peoples based on the experience of hunter-gatherers from Australia. At the time of European contact, the continent was populated by about 300,000 people without a single domesticated plant. Food supply was largely by plant gathering not by the hunting of game. Early humanity, from the beginning, depended upon botanical knowledge for existence. They became familiar with literally hundreds of species. They knew how to clear or alter vegetation with fire, sow seed, plant tubers, and protect plants. They laid claim to individual trees and tracts of land, celebrated first-fruit ceremonies, prayed for rain, and petitioned for increased yield and abundant harvest. They spun fibers, wove cloth, and made string, cord, baskets, canoes, shields, spears, bows and arrows, and a variety of household utensils. They painted pictures, carved masks and ritual objects, recited poetry, played musical instruments, sang, chanted, performed dances, and memorized legends. They harvested grass seeds, threshed, winnowed, and ground them into flour. They dug roots and tubers. They detoxified poisonous plants for food and extracted poisons to stun fish or kill game. They were familiar with a variety of drug and medicinal plants. They understood the life cycles of plants, knew the seasons of the year, and when and where the natural plant food resources could be harvested in greatest abundance with the least effort. As long as populations were below the carrying capacity of the land, famine and starvation were rare. There was a large amount of leisure time. In some ways, it was the golden age!

Man as hunter is another aspect of the human experience that has had a profound impact on human culture. Prehistoric artifacts and Neolithic cave paintings provide a clear insight into the origins of hunting and hunting technology via remains and pictorial representations of weapons (the spear, bow and arrow, the missile), evidence of hunting organization, and depictions suggesting religious and mythic influences. In the once smoky caves, we see dazzling pictures of animals with little if any suggestion of plants. Yet, Paleolithic burials show high pollen counts that suggest evidence of floral offerings to the dead.

The hunting experience has not disappeared from our present-day culture. It remains in the suburban fox hunt—"the unspeakable chasing the inedible"—the lure of the National Rifle Association, in the participation of weekend warriors in war games, the longing to cover our bodies with fur and leather, and our love of hunting canines now transformed into spaniels and lap dogs. The domestication of animals modified hunting into herding and nomadic life. The symbiotic relation of people and grazing animals (sheep, goats, cattle, camels) with pasture and forage plants created a tripartite bond that is reflected in the metaphors of the good shepherd and the cowboy. Herding took people out of the cave and placed them under the stars during their trek for green pastures and provided time for contemplation, to develop soaring poetry, religious sagas, and the beginnings of astronomy/astrology.

## THE AGRICULTURAL REVOLUTION

The most sweeping technological change for humanity occurred in prehistory: the use of tools, the discovery of fire, and the invention of agriculture. Knowledge concerning the origins of agriculture comes to us from the diggings of archaeologists but also from deep, subconscious memories that surface in the form of legend, myth, and religious teachings. Agriculture appears in the compressed time warp of the archaeologist as a sudden transformation referred to as the Neolithic Revolution. About 10,000–12,000 years ago, in the highlands of the Tigris-Euphrates River complex, there appeared the sure remains of sedentary existence and the beginnings of cultivation. The history of humankind irrevocably changed.

The precise origins of agriculture are unknown. The traditional view has been that agriculture is a divine gift. In Egypt, the source was the goddess Isis, the black soil of the delta, created by the touch of her husband and brother, Osiris, the Nile itself. It was she who discovered wheat and barley and, in thanksgiving, was worshipped in December, when the sun was reborn, as she nursed, in a stable, her miraculously conceived son. In Greece, she was Demeter and, in Rome, she was Ceres and Flora; in the New World, agriculture is a gift of Quetzalcoatl, a god disguised as a plumed serpent.

The biblical account in Genesis is the other side of the coin. Agriculture is not a gift, but a curse. Adam (man) was originally given every herb-bearing seed and was provided abundant provision in the garden of Eden. Eve's quest for knowledge is punished by the pain of childbirth and subjugation to her husband, while Adam must toil by the sweat of his brow. Their children continue to get the same message. The offering of Abel (the nomad) is preferred over Cain's (the cultivator), and the first conflict between rancher and farmer is decided violently. (Clearly, a nomad, with time to contemplate the stars and to chant soaring poetry, was the author of the book of Genesis.)

The universally accepted theory for the origin of agriculture is that it developed from a series of inventions emanating from the discovery, by sedentary people, of seed or vegetative propagation. Cultivated plants and domesticated animals substituted for the bounty of wild species previously harvested by gathering and hunting. This concept was first formulated by Charles Darwin, and the details were modified by various students of agricultural beginnings. Jack Harlan (1975) has made the point that every model for the origin of agriculture has evidence against it and concludes that the best model is no model. His conclusion is that humankind found various reasons and various ways to come up with similar solutions to fundamental problems. Whatever the precise impetus for the origins of agriculture, there is no question that the effect has been cataclysmic.

Survival systems based on gathering and hunting are successful only when the human population is maintained well below the carrying capacity of the land. Harlan has made a good case that present-day hunting societies function with little effort and great leisure and desist from agriculture more out of inclination than lack of knowledge or skill. Despite the leisure and ease of present-day remnants of gathering and hunting societies, these cultures are restricted and cannot expand. In fact, it is clear that all hunting societies developed strategies to restrict population by a combination of tactics such as warfare, sexual codes to delay conception or to restrict fertility, and perhaps murder, be it euthanasia for the old or infanticide for the young. Whatever the precise impetus for the exchange of the vagaries and thrill of the hunt for the security and toil of agriculture, a way was found to maximize productivity by domestication of wild plants and animals. Agriculture ensures increasing food production, but it also makes high population an economic asset for the producer. Thus, modern mores that extol the sanctity of life, high fertility, and the work ethic are a direct result of the agricultural revolution.

The discovery of agriculture is remarkable for two reasons. One is its universality. Although it is tempting to ascribe a single locus and a diffusion pattern, the evidence suggests that agriculture was an independent discovery throughout many parts of the world. For example, each great ancient civilization was based on grain—a nutritious, compact, and versatile source of food: wheat in Europe and the near East, rice in Asia, maize in the Americas, and sorghum in Africa—thus demonstrating the technological brotherhood of humankind. The second remarkable aspect of agriculture is the perspicacity of each population in ferreting out the desirable species—plant and animal—and transforming them into new entities—crops and livestock, a process known as domestication.

Domestication involves two distinct events. One is to identify potentially useful species and the other is actually to transform them into dependable servants. The latter is accomplished by no less than a genetic transformation achieved through selection of genetic variants that intensify desirable traits and eliminate undesirable characteristics.

The choice of appropriate species seems obvious when it is completed, but so are all acts of genius. The virtue of the original, unimproved, selected species may not have been so obvious. Cassava, for example, is poisonous, and many crops are unpalatable or inedible without the cooking process.

Selection (differential reproduction) led inexorably to evolutionary changes, as the process converted weedy plants to new species dependent upon people to complete their life cycles. The traits desirable from a human perspective are often ones that prevent survival of the plant. Cultivated plants, unlike weeds, are usually unadapted to exist without the benefit of human interference. The development of crops resulted in a loss of independence of both people and plants. As in the case of the dairy farmer and his herd, it is not clear who serves whom the most. Many crops, maize, for example, have been so altered that they no longer exist outside cultivation, and a direct connection to their ancestors has been obliterated.

The success of domestication assured the expansion of agriculture. Examples of fundamental alterations in crops are changes that ensure dependable cultivation and increase harvestability (e.g. loss of seed dormancy and seed shattering) and alterations that increase productivity, usually by altering the proportion of the plant that is economically useful (harvest index) rather than an increase in true biological efficiency.

The end result of the agricultural revolution has been a fundamental change in the human condition. The interaction of people, crops, and domestic animals has resulted in fused genetic destinies. An abundance of food causes changes in selection pressure and alterations of human evolution equivalent to those wrought by the domestication of plant and animal species.

Agriculture, by creating not only a dependable food supply but a surplus to be stored, permitted civilization to develop. In the process this new system pushed out the hunter and the nomad and rapidly expanded to all usable land, filling it with people even beyond its capacity! As agriculture produced more food, it instilled the quest for fertility—of corn, of cattle, of soil. The present population explosion has its roots in that phenomenon.

The social ramifications of the Neolithic Revolution remain. They include the implications of territoriality, our feelings regarding fertility and population, and our attitude regarding community. Important technological questions remain unresolved. Have our forebears made, in fact, the best choices of servant species? Are we hostage to the solutions of our forebears, or can we begin anew? One is awed by the conservatism of the human species seemingly held captive by the resource base of the past.

## THE RISE OF HORTICULTURE

The intensive use of plants for people is the meaning of horticulture. Its origins are inseparable from the beginnings of agriculture. Horticultural art and technology survived the dark ages in monastic gardens and gardening became an integral part of monastic life, providing food, ornament, and medicines. The gardener (hortulanus or gardinarius) became a regular office of the monastery. As feudalism gave way to trade, garden culture was taken over by secular rulers and the rising monied middle class. The rise of horticulture during the Renaissance is to be found in the formal gardens and magnificent landscapes of European royalty, and reached a peak in the Palace of Versailles, constructed over a 25-year period by Louis XIV and designed by his architect, Andre Le Nôtre (1613–1700).

The resurgence of horticulture comes to us in the illustrated and printed herbals, in the threads of medieval tapestry, and romantic poetry. Shakespeare, the greatest writer in English—if not the greatest in any tongue—is a rich source of horticultural information. The Renaissance came late to England but flowered with a brilliance that still interests humanists and scientists alike.

The great bulk of Shakespeare's imagery is drawn from everyday things, seen and unseen. Of all the images of nature, the greatest number is devoted to horticulture. Who can forget that "rough winds do shake the darling buds of May." Shakespeare displays an intimate knowledge of plant growth, propagation, grafting, pruning, manuring, weeding, ripening, and decay. Over 200 plants are mentioned. The horticultural imagery and allusions have become part of our culture and fuse plant lore and literature. The gardener to the King in Richard II is so instructed:

> Go bind thou up young dangling apricocks,
> Which like unruly children make their sire
> Stoop with oppression of their prodigal weight;
> Give some supportance to the bending twigs.
> Go thou, and like an executioner
> Cut off the heads of too fast growing sprays.
> That look too lofty in our commonwealth;
> All must be even in our government.
>
> SHAKESPEARE, RICHARD II (III, iv, 29–36)

## PLANTS AS SYMBOLS

Symbols invoke images deeply embedded in the human experience. In many cases, the original meanings have become changed and subverted as layers of new meanings are superimposed over the old. The significance of plants to human history—for sustenance as well as healing—has created a rich treasure of plant symbolism that we cherish without a clear understanding of its origins, yet its power is such that we pass it on to our children generation after generation. This has intensified with the rise of agriculture and it is no accident that the symbolism of sexuality and fertility is especially powerful in plants—scarcely repressed in present-day romantic uses of flowers.

Thus, the poisonous white fruit of the parasitic European mistletoe, *Viscum album*, was once considered to be the generative power of the oak god. In the Renaissance, remembering its phallic significance, mistletoe became a love charm, and this memory is retained in our modern custom of kissing under the mistletoe at the Christmas season.

Flowers still retain extensive ceremonial use in the expression of joy, affection, welcome, gratitude, sympathy, celebration, grief, friendship, marital union, or spiritual contemplation. Flowers have become the "language of love," and we still "say it with flowers."

Plants with strong aromas and flavors very early attracted human attention, probably, at first, for magical rites, spells, purification ceremonies, and embalming, as well as for fragrances, perfumes, and cosmetics. We prefer their odor to our own as we sweeten our breath with their essences and perfume our bodies with their scents. Fragrances and perfumes have always played enticing roles in love making. The Song of Solomon combines eroticism with metaphors of spices and vegetative fertility. It is indeed a miracle that, through convoluted exegeses, the Song of Solomon has survived as a portion of Holy Scriptures.

The ancient traffic in spices is celebrated in the story of Joseph and his brother, who "saw a caravan of Ishmaelites coming from Gilead with their camels bearing gums, balm, and myrrh on their way to carry it down into Egypt." The voyage of Columbus, whose 500-year anniversary we soon celebrate, was lured by a shortcut passage to the spice-rich Indies. The new crops discovered in the New World were to prove more valuable than all the gold and silver stolen from the treasuries of the New World empires.

Towering over mortals, trees have always been powerful symbols endowed with spiritual meaning. They have long been worshipped, some times as oracular gods or goddesses. Many myths refer to the Tree of Life, and the sacredness of trees is retained as we knock on wood for luck. The story of Eve and the forbidden fruit (sometimes considered a fig, but transformed to an apple in our time) and the Christmas tree, which dates to pagan ceremonies in Southern Europe, retain the emotional impact of tree symbols.

The importance of plants to human history has made them powerful national symbols. The fleur-de-lis symbolizes France, the shamrock Ireland, the leek Wales. The red and white Tudor rose is the royal emblem of England, symbolizing the politic marriage of Henry Tudor to the Plantagenet Elizabeth of York, which cemented the peace after the Wars of the Roses, 1455 to 1485, between the House of Lancaster, which chose the red rose as its symbol, and the House of York, which carried the white rose.

Life cycle events are still memorialized by plants. Cigars are given out at our birth, and our first date and sexual awakening is ritualized in the exchange of floral sex organs—the corsage. Our fiancees are swept off their feet with roses, and the marriage couple is showered with rice to bestow fertility. We celebrate motherhood with flower arrangements, and, when gravely ill or near death, some are anointed with blessed olive oil; our deaths are memorialized with floral wreaths.

Our holidays—both secular and religious—are rich in plant symbolism. We toast the New Year with champagne, associate the father of our country with a cherry tree and the great Emancipator with a log cabin or split rails. On Valentine's Day, the florists get well. The ashes of burned palm fronds are daubed on Catholic foreheads on Ash Wednesday, and their lapels are graced by crosses of fronds on Palm Sunday.

The Jewish festival of Passover is celebrated with unleavened bread, bitter herbs, and a mixture of apples and wine. The wine and bread received during the celebration of mass by Christians transubstantiate to the blood and flesh of the Son of God. We remember our veterans with poppies, and our children celebrate Halloween with jack-o'-lanterns made from pumpkins. Thanksgiving is unthinkable without cranberries, and Christmas is synonymous with evergreen trees and wreaths.

Plants are also symbols of evil. Some plants have been, and will continue to be, associated with the dark side of nature. Shakespeare is obsessed with weeds as metaphors for decay and death. Hamlet declaims:

> How weary, stale, flat, and unprofitable
> Seems to me all the uses of this world!
> Fye on it, ah fye! 'tis an unweeded garden
> That grows to seed. Things rank and gross in nature
> Possess it merely.
>
> HAMLET (I, ii, 133–137)

The Queen in 2 Henry VI observes:

> Now 'tis the spring, and weeds are shallow-rooted;
> Suffer them now and they'll o'ergrow the garden,
> And choke the herbs for want of husbandry.
>
> 2 HENRY VI (III, i, 31–33)

In our time, we have killer tomatoes, kudzu, and coca, which produces cocaine and crack, not to mention the tobacco in Kools, Carletons, and Camels. We also have stinging nettles, thorny brambles, poison ivy, poison oak, poisonous mushrooms, hay fever plants, crabgrass, and, my movie favorite, the odoriferous murky bogs where the villain quickly sinks up to his neck and slowly, horribly, subsides screaming, as he disappears below the ooze. Finally, George Bush has tried, with some success, to add broccoli to the list of evil plants.

## PLANTS AS CURES

Herb lore was a complex discipline in primitive society, and the accumulated wisdom was dispensed by practitioners who became sacred healers with great power and prestige, despite their inability to cure in most cases—a tradition carried over to our modern physicians. Knowledge of the curative power of plants is the beginning of medical wisdom. The accumulation of knowledge, as well as misinformation, is stored in ancient and modern herbals that, to this day, have not lost their appeal. Probably, as a result of disaffection with the complexity of late 20th century life, herbal lore has undergone a resurgence as we sip herbal teas, gorge on oat bran cereals, and eagerly take up skin rejuvenators and elixirs. Yet, many of our most important medicines originate from plants: aspirin from willow bark, reserpine from Indian snakeroot (*Rauwolfia serpentine*), colchicine from autumn crocus, quinine from cinchona. The history of drugs is largely the early history of botanical science. The famed herbal of Dioscorides in the 1st century became the sourcebook for physicians for 1500 years!

Over 10% of dry mass of some plants is made up of chemicals designed for defense against predators. What is one creature's poison is another creature's medicine. The screening of plants for new efficacious pharmaceuticals is still in its infancy. It has been estimated that only 5,000 of 300,000 species have been intensively investigated. Plants are some of the most complicated chemical factories known, and we risk our future health if we ignore them or, worse, let them disappear through extinction.

## THE INDUSTRIAL REVOLUTION

The rise of the industrial revolution in the 19th century led to the industrialization of agriculture—first by mechanization and then by the application of chemistry in the form of fertilizers, herbicides, and pesticides and biology via genetics, plant breeding, and biotechnology. This has produced a population explosion in the developing world and a surplus of food in the developed world at the same time it has resulted in a dearth of farmers. In the United States, there are more unemployed workers than agricultural producers. In the rich countries of North America, Europe, and the Pacific Rim, the typical person is on the verge of being cut off from direct contact with plant life. Our athletic fields have been stripped of grass and replaced with Astroturf; our cotton jackets are made of nylon; our bouquets are silk flowers; our once wooden school desk is now plastic. Even our great parks are being strangled, with the light cut off by skyscrapers and open spaces eaten up by buildings and roads. Worse, the crime rate of our cities has become such that our parks become repositories of fear as soon as the sun goes down.

The changes in agriculture brought about by the industrial revolution brought into sharp focus many problems with which traditional practices could not cope. Progress in agriculture had been, up to the early 19th century, largely derived from empirical studies. The accumulated successes and improvements had become embedded in human consciousness via legend, craft secrets, and folk wisdom, and had become part of our modern culture. This information was stored in tales, almanacs, herbals, and histories. More than practice and skills were involved as improved germplasm was selected and preserved via seed and graft from harvest to harvest and generation to generation. The sum total of these technologies made up the traditional agricultural lore.

The scientific tradition is not as old as the empirical but it is ancient nevertheless. Its beginnings derived from attempts at systematic discovery of rational explanations for nature. Science, from the Latin "to know," is a method of accumulating new information about our universe. The driving imperative is the desire to understand. If necessity is the mother of invention, curiosity is the mother of science. The scientific method involves experimentation, systematic rationality, inductive reasoning, and constant reformulation of hypotheses to incorporate new facts. When new explanations of natural phenomenon are accepted, they must nevertheless be considered not as dogma but as tentative approaches to the truth and, as such, subject to change. The process is cumulative, and science is only alive when it grows. When any society claims to know the complete truth and that further questioning is heresy, science dies.

The rise of modern science in botany and agriculture has led to a scientific establishment in public and private universities and state and federal experiment stations. These institutions have developed a vast pool of agricultural information on soils, diseases, and insects and have developed improved cultivars of plants that provide the basis for increased agricultural productivity. The enormous advances made in the agriculture of the developed world is a testimony to the correctness of this approach. Our present world food problem is a problem of poverty and politics—not a lack of information or know-how regarding agriculture.

## THE FUTURE

We are now entering the future—the post-industrial era. It is a world of instant communication, supersonic travel; it is also the world of the global village, ever-expanding population, and pollution. The post-industrial era traces its beginnings to the 1960s with three crises impinging on world consciousness. The first was apprehension of an emerging world food shortage caused by unchecked population increase in the tropical world. The second crisis, first brought into prominence by Rachel Carson's *Silent Spring* (1962), was the pollution resulting from technology itself. The third was the energy crisis of the 1970s, when the 1973 Arab oil embargo caused a quadrupling of crude oil prices and initiated a general economic disorientation. These three problems were soon shown to be part of a single issue—the distribution of the earth's renewable and nonrenewable resources. Ironically, as we raced to the moon, we rediscovered the earth. Whereas our plant-dominated planet had traditionally been thought of as "Mother Earth," the source of bounty and plenty, the new symbol of a "Spaceship Earth" stressed the finiteness of earth's resources. The new earth was seen as if from the vantage of an observer on the moon: small and isolated yet beautiful and precious as never before.

The food crisis was put in abeyance with the Green Revolution and a slow-down of population increase and the oil crisis self-destructed by the greed of the oil producers—postponed for another day. But the ecology movement has grown and become a political force in the United States and Europe. The "greens" are a power in Europe and are becoming an independent political force in the United States. We have previously identified green with the color of money but now the color green is taking on an entirely new meaning, derived from the color of chlorophyll. Conceived in anxiety and born in fear, the ecology movement inspired religious fervor and struck a responsive chord in the United States in the late 1960s and early 1970s. Although its message was strident and its thrust sometimes veered uncomfortably in an anti-intellectual, antiscience direction, the ecology movement has proved to be a force to be reckoned with. The banning of DDT, its cause célèbre, was the most notable of its victories. The ecology movement turned attention to social problems and displayed concern for the displaced agricultural worker, the human jetsam of agricultural progress. Many of the causes of the 1990s—to save the rain forest, to reduce acid rain, to slow the increase of atmospheric $CO_2$ and global warming, to prevent a widening in the ozone hole—are largely concerned with plant–people interactions. We have begun to think of the green plant as the canary in the mine—an early warning against deadly peril.

As our space program is renewed and girds again for the challenge of the heavens, the notion of establishing space colonies has become less a dream than a planned goal. It has become obvious that there is not room enough in our space craft for our sustained requirements for oxygen and food. The thought of jettisoning our wastes to float forever in space is intolerable—all too reminiscent of the Muppets' "Pigs in Space." Rather, we must recycle our wastes, purify our air, and provide sustenance by using sunlight as an outside energy resource. The machine that our engineers have devised that will marvelously recycle our wastes, release oxygen, consume carbon dioxide, and provide our sustenance in the form of carbohydrates, proteins, and lipids is already found in a handful of magical bean seeds. In short, space engineers have rediscovered plants. The green plant, so plentiful and readily available on earth, will prove to be our savior in space. Horticulture again—now in the guise of space biology—has reestablished the essential connection between plants and people. Horticulture must increase in importance, in schools, in homes, in communities, to underscore the interconnectedness of the living world and to improve the beauty and the quality of life here on earth.

## *BIBLIOGRAPHY*

Berrall, J. S. 1966. The garden: An illustrated history. Viking Press, New York.

Clayton, V. T. 1990. Gardens on paper. Prints and drawings 1200–1900. University of Pennsylvania Press, Philadelphia.

Coats, P. 1970. Flowers in history. Viking Press, New York.

Ellacombe, H. M. 1896. The plant lore and garden craft of Shakespeare. [Reprinted 1973 AMS Press, New York.]

Gothein, M. L. 1913 (German edition); 1925 (Second German edition); 1928 (English edition); 1966 (Reprint of English edition). A history of garden art. Hacker Art Books, New York (2 volumes).

Harlan, J. R. 1975. Crops and man. American Society of Agronomy, Crop Science Society of America, Madison, Wisconsin.

Huxley, A. 1978. An illustrated history of gardening. Paddington Press, New York and London.

Hyams, E. 1971. A history of gardens and gardening. Praeger, New York.

Janick, J. 1979. Horticulture's ancient roots. HortScience 14:299–313.

Janick, J. 1986. Horticultural science (4th ed.). W. H. Freeman, New York.

Janick, J. (ed.). 1989. Classical papers in horticultural science. Prentice-Hall, Englewood Cliffs, New Jersey.

Janick, J. 1990. Agricultural revolutions and crop improvement. In: P. V. Ammirato, D. A. Evans, W. R. Sharp, Y. P. S. Bajaj (eds.). Handbook on plant cell culture, Vol. 5. McGraw-Hill, New York.

Janick, J., R. W. Schery, F. W. Woods, and V. W. Ruttan. 1981. Plant science: An introduction to world crops (3rd ed.). W. H. Freeman, New York.

Lees, C. B. 1970. Gardens, plants, and man. Prentice-Hall, Englewood Cliffs, New Jersey.

Leonard, J. N. 1973. The first farmers. Time-Life Books, New York.

Moldenke, H. N. and A. L. Moldenke. 1952. Plants of the bible. Chronica Botanica, Waltham, Massachusetts

Walker, B. G. 1988. The woman's dictionary of symbols and sacred objects. Harper & Row, San Francisco.

Wright, R. 1934. The story of gardening. Dodd, Mead & Company, Garden City, New York.

CHAPTER 2

# *The Corporate Garden*

Dana C. Parker

Longwood Graduate Fellow

## *INTRODUCTION*

Corporations in America have increasingly supported the arts and cultural activities in their communities and throughout the country. Such support is motivated by self-interest and financial incentives in the form of tax deductions. In the area of horticulture, this corporate stance is demonstrated by well-groomed grounds, public access, and the attitude that gardens and park-like settings enhance the corporate image. The landscapes include gardens that improve the workplace and provide benefits to the community and the corporation. Corporations have, with gardens, the resources and opportunity to patronize the arts, extend their philanthropy, and reflect a positive image in their communities.

Little research has been conducted on the work environment, but its importance is underlined by the amount of time spent at work. The workplace is now being linked to psychological needs, performance, and well-being. "The workplace becomes a kind of substitute public location, one of the only places where appreciable numbers of people actually come face to face on a regular basis, where the better part of waking hours is spent in close proximity to one another" (Koetter, 1987).

Some corporations are reappraising their approach to the people on which their profits rely, both staff and the public. The provision of artwork, plantings, and physical fitness opportunities are not just efficiency measures, but recognition that company success is based on mutual respect. Gardens are part of the concern for the office environment and every office building says something about the organization it houses. From an architectural point of view,

> a building can create a corporate identity by enhancing its immediate surroundings by contributing to the urban or suburban scene. The strongest image is not always visual, but may also be psychological and respond to providing amenities and good neighborliness. Amenities can take the form of plazas, atria, gardens, trees, works of art, or simply a place to sit (Marquis, 1970).

## DEFINITIONS

A garden is "a place for public enjoyment planted with trees, flowers, etc., and often having special displays of plant life" (Webster's, 1983). The addition of "corporate" simply places the garden in a distinctive, business setting. Public enjoyment assumes availability and accessibility to corporate properties by the public with no entrance fees such as might be associated with theme parks or other corporate-based landscapes.

To consider gardens as an expression of philanthropy, the definition goes beyond cash contribution, which is often used to measure philanthropy. This expanded notion of philanthropy is "the spirit of active goodwill toward one's fellow men, especially as shown in efforts to promote their welfare, usually demonstrated by gifts, institutions, services, or acts" (Webster's, 1942).

Corporations in this study were limited to those not deriving profits from the landscape or horticulture industry. The selection of the corporate headquarters building became a natural focus because this is where the gardens were found. This building is the most important as the corporate symbol of image and prestige, and consequently where the corporation is willing to spend extra time, money, and effort. As the location of top management, the image of "corporate headquarters" is very important.

Corporate gardens are found in urban and suburban locations. In the urban context, the gardens can be found in plazas, parks, and roof tops. The atrium has become a generic building form of both urban and suburban architecture that provides space for indoor gardens. The suburbanization of America has witnessed the phenomenon of the nonurban workplace emerging commonly as suburban office parks and campus-type environments, which some have dubbed "corporate villas" (Goldberger, 1983).

## THE CORPORATION

Corporate image is much lower today than it was twenty years ago, as news reports increasingly focus on oil spills, product recalls, take-overs, mergers, lay-offs, and strikes. To counteract negative opinion, corporations are realizing the importance of image in the feeling it creates with its products, business dealings, relations with community, and the appearance of its properties.

The increasing importance of top management's values and interests was a trend that surfaced in research of corporate gardens. The Chief Executive Officer (CEO) is the most important figure in determining the corporate culture, which is the shared values or way things are done. Attention to aesthetics, amenities, social responsibility, and employee comforts may all be tangible aspects of corporate culture. Thus, gardens can be a tangible aspect of "corporate culture."

Corporate patronage of the arts and corporate philanthropy were found to be parallel and overlapping interests. The goals of both forms of public relations are largely the same; the last two goals are more specific to art collecting:

1. Improving the environment in which to work and do business.
2. Improving the local community.
3. Improving public relations.
4. Increasing profitability with the ability to recruit quality employees.
5. Supporting local artists.
6. Establishing the status and value of the collection.

Corporate philanthropy has historically and legally been linked to self-interest (Karl, 1982). A 1982 survey of corporate executives by the Council on Foundations confirmed ideas about corporate giving (White and Bartolomeo, 1982). This survey found that the primary agenda for corporate giving is self-interest and that the CEO is the most influential person regarding giving decisions. Many corporations realize that a rich cultural environment improves the community. As patrons of the arts, corporations can improve the quality of life for their public and provide valuable support for creativity, aesthetic environments, and opportunities for the arts.

## ROLE OF THE GARDEN

Historically, the great estate gardens of the past have symbolized wealth, power, and prestige and are considered an art form. The garden is a symbol of success. Other roles for the corporate garden are aesthetic enhancement, amenity or service, public relations tool, educational or cultural asset, and recreational or social setting.

Corporations do not support the arts or gardens for strictly altruistic goals. They need justification that supports self-interest. Although difficult to quantify, benefits lie in the area of employee considerations, financial incentives, recognition and awards, and improved public relations.

## BENEFITS TO THE CORPORATION

The economics of the office building attributes 90% of costs to employees (salaries and benefits). Only 10% is for the creation, construction, and operation of the building (Goodrich, 1982). The physical environment affects employees' ability and desire to work. Health cost and fitness considerations are placing new importance on the office landscape as a resource for active and passive enjoyment. Research by an anthropologist at an office park in New Jersey showed that people place a high value on the beauty of the landscape and the recreational or social activities that it affords (S. Low, unpublished).

Employee benefits and amenities are a way to attract and retain high-quality employees. Both the Codex Corporation and the John Deere Company point to high-quality design and the landscape as their most important recruiting tool. The Deere atrium is noted as one of the best of its kind for enhancing the work environment (Saxon, 1987). Management noticed an increase in productivity, morale, and pride in the workplace.

Another financial consideration can be lower taxable income; the costs and maintenance associated with gardens are legitimate business expenses that can be deducted. Corporations can also deduct charitable contributions to lower their taxable income. The Rhododendron Species Foundation on the property of the Weyerhauser Corporation is a not-for-profit foundation that receives land, funds, and facilities from the corporation. This is an unusual and innovative partnership that provides benefits for the corporation and the foundation.

Zoning regulations can have financial implications as many cities encourage private development of public spaces such as atria, plazas, and parks. New York City led the way in this type of zoning legislation in allowing IBM to build the Bamboo Court in exchange for additional height, thereby allowing increased floor space and obvious financial gain.

Construction of atriums is now considered to provide economic returns, because atriums can be relatively inexpensive to build and can recycle older buildings, provide increased earning power, and raise productivity (Saxon, 1987). Property value and increased earning power are important incentives when the addition of gardens can command higher rents and occupancy.

In public relations, the landscape can be considered another channel of communication. Corporate gardens can be a community resource for recreation, education, and cultural opportunities. Many corporations provide access and a variety of opportunities in their gardens that make meaningful contributions to the community.

## CONCLUSION

"Public art's best chance in this age of corporate and bureaucratic hold on public experience may lie in intimacy in providing an oasis, a garden, a home in the vastness and impersonality of public contexts" (Lippard, 1981). Corporations can support the arts and philanthropy with gardens. Some corporations recognize this by investing in their communities and providing amenities and patronage to their interests. Gardens in the workplace can be a means to accomplish profit, serve human needs, and provide an element of social responsibility. Through a collaboration of the arts and business, corporations can improve the aesthetic environment in which we work and live.

## LITERATURE CITED

Goldberger, P. 1983. A corporate equivalent of the classical villa. New York Times. 3 July 1983.

Goodrich, R. 1982. Seven office evaluations. Environment and Behavior 14:354.

Karl, B. D. 1982. Corporate philanthropy: Historical background. In: Corporate philanthropy: Philosophy, management, trends, future, and background. Council on Foundations, Washington, D.C.

Koetter, F. 1987. The corporate villa. Design Quarterly 135:6.

Lippard, L. 1981. Gardens: Some metaphors for a public art. Art in America 69:136.

Low, S. Working landscapes: A report on the social uses of outside space in corporate centers and program recommendations for Carnegie Center. Unpublished report.

Marquis, H. H. 1970. The changing corporate image. American Management Association, Washington, D.C.

Saxon, R. 1987. Atrium buildings: Design and development. Van Nostrand Reinhold Company, New York.

1983. Webster's new twentieth century dictionary. Prentice Hall, New York.

1942. Webster's collegiate dictionary. Springfield, Miss.

White, A. and J. Bartolomeo. 1982. Corporate giving: The views of chief executive officers of major American corporations. Council on Foundations, Washington, D.C.

CHAPTER 3

# Gardens and Civic Virtue in the Italian Renaissance[1]

Lawrence W. Rosenfield

Professor of Communication Arts, Queens College, City University of New York

Scholarship detailing the place and practice of rhetoric in the Italian Renaissance has recently flourished.[2] That rhetoric occupied a central position seems beyond dispute (Gray, 1963; Hardison, 1971; DeNeef, 1973; Lee, 1967). More problematic has been its role in reconciling the *vita activa* and the *vita contemplativa*, the life of civic activity and the withdrawn pursuit of private, philosophical endeavors. As Pocock (1975) remarks,

> The humanist was ambivalent as between action and contemplation; it was his *metier* as an intellectual to be so, and he could practice it perfectly well within his framework of the republic.

This duality produced neither paradox nor tension for the Renaissance citizen. On the contrary, the two elements were indispensable to each other. In the short time available today, I shall sketch that rhetorical filigree linking action and contemplation with the illustrations drawn from a prominent epideictic vehicle—the Renaissance garden.

The classical world had taken for granted that life in a *polis* or republic was fulfilled in direct association with others, in conversation. It is also recognized that such activity was fatiguing, that rhetorical conduct depleted the individual, and deprived him of those "virtues" needed for the discharge of public affairs (Pocock, 1975). The aim of civic humanism was to be fully engaged in the body politic. Such a condition called for two distinct rhetorical activities: thought and discourse to carry on public business, and thought and discourse to celebrate or acknowledge appreciation for the human being. The latter task fell to epideictic.

Thomas Wilson (1963) echoed the traditional view concerning the scope and purpose of epideictic; he claimed for eloquence the ultimate goal of *civilizing* the members of the body politic.

Whereas men lived brutishly in open fields, having neither house to shroud them in, nor attire to clothe their bodies, nor yet any regard to see their best avail, these [orators] appointed by God called them together by utterance of speech and persuaded them what was good, what was bad, and what was gainful for mankind. And . . . being somewhat drawn with the pleasantness of reason, and the sweetness of utterance . . . after a certain space they became . . . of wild, sober; of cruel, gentle; of fools, wise; and of beasts, men; such force hath the tongue and such is the power of eloquence and reason, that most men are forced, even to yield in that which most standeth against their will.

So epideictic stands as a precondition for other rhetorical activity. It serves to shape and sustain civil society, to dissolve bellicosity, obduracy, and fickleness, to rehabilitate the populace into a more noble state than their lowborn natures might ordinarily allow. It tries to nurture in the individual those traits of good citizenship needed for optimal service to the principality. The Italian Renaissance enriched ceremonial theory and practice with three ideas represented in the works of Boccaccio, Castiglione, and Machiavelli. Although these authors did not address rhetorical theory as such, they are convenient spokesmen for emphases displayed in the age.

Boccaccio advocates the solitude needed by the man of letters, the opportunity for meditative repose to restore both poet and orator so each may return refreshed to the bustle of public interaction. In his essay, "On Poetry," he notes, with characteristic feistiness,

> Poets sing their songs in retirement; lawyers wrangle noisily in the courts amid the crowd and bustle of the market. Poets long for glory and high fame; lawyers for gold. Poets delight in the stillness and solitude of the country; lawyers in office buildings, courts, and the clamor of litigants.
> Furthermore, places of retirement, the lovely handiwork of Nature herself, are favorable to poetry, as well as peace of mind and desire for worldly glory; the ardent period of life has very often been of great advantage. If these conditions fail, the power of creative genius frequently grows dull and sluggish.

and later in the same essay,

> The wordsmith escapes to rural climes not to live as a hermit, but to free himself from distraction and to replenish those creative reserves which are inevitably drained by the hectic social intercourse of court and city (Osgood, 1956).

Castiglione represents the orator's allegiance to decorum, his ability to adapt his discourse to the needs of audience and circumstance in order to accomplish his persuasive aims. Rhetoric's paradox is that it is only by respecting his audience's expectations that the orator can hope to change the audience. Hence, the tacit rules of custom and usage that are familiar to a community are the foundation for influencing that community. The flexibility of rhetorical invention demanded by decorum has as its ethical counterpart the virtue of prudence, the capacity to fuse reason and action in ways appropriate to the circumstances (Kahn, 1983).

From Machiavelli comes a final Renaissance rhetorical precept: bold action and the capacity for crafty, unconventional thinking presume the detachment from commonplaces, a mental flexibility that comes from widened perspective. His observation in *The Discourses* that "the great majority of mankind are satisfied with appearances" (Ricci, 1940) does not counsel insincerity; it only voices the Renaissance appreciation of the *visible*. What appears in the world bears no necessary relation to the private motives that lurk in men's hearts. The membrane separating the domains of thought and conduct is self consciousness, the momentary hesitance besetting one about to commit himself irrevocably by acting into the world and the laws of visibility that govern the world. The gap between the two realms is bridged by one's *virtu*, the confidence in one's verbal, political, and martial skills that grows from one's detachment.

Brevity compels over-simplification. Epideictic civilized men; ceremony restored the individual to a condition in which he was fit to conduct public affairs with prudence. One opportunity for restoration in the Renaissance came in moments of quiet reflection in "natural" settings. Were such contexts embellished with reminders of the common sense and decorum undergirding the republic, their epideictic worth was enhanced. Were the settings further designed to restore the sense of perspective—that detachment from immediate events—so that the individual's *virtu* could reassert itself with appropriate clarity, the setting's value as an epideictic vehicle was enriched even more.

Of course, many ritual occasions met the general ceremonial needs of commemoration, but the Renaissance produced one unique state for achieving epideictic aims, the "pleasure ground" of the courtly garden. I turn in the time remaining to suggestion of a few of the rhetorical elements that were incorporated into the new art of garden design in order to help it accomplish its civilizing task.

The garden's rejuvenating function is apparent. For example, the rhetorician Pico della Mirandola astutely remarked on the good fortune of the man who could repair the breach between physical and spiritual selves by temporarily setting aside public business to dwell in a holy shrine in the midst of sacred groves.

> Then, as these [sylvan] pleasures possess both the eye and the ear, they soothe the soul; then they collect the scattered energies of the mind, and renew the power of the poet's genius . . . (Comito, 1978).

This fabricated earth did more than relax. It also symbolized a fitting internal landscape for the garden dweller. It did this, according to Pico, by using an ensemble of icons that operate on the soul in the same fashion as does the sweetness of effective rhetoric (Comito, 1978).

> The garden models the way in which the mind conceives its relation to the world external to itself; but gardens . . . become arenas in which the externality of the world is at least temporarily overcome, a truce proclaimed in the continual battle between the shows of things and the desires of the mind . . . (Comito, 1978).

To this end of establishing a correspondence between the tranquility of the garden and the tranquility of the inner self, a stroll through the garden landscape became a metaphysical journey of instruction in how one might, by "imitating" the garden terrain, achieve a comparable inner joy and satisfaction (Cool, 1981).

The garden's "message" was in the main epideictic. The visitor came to the garden as a spectator, to celebrate the glories of human being. His object was to behold and take to heart the delights the garden disclosed. And what was revealed in both the public and private spaces alike was the epideictic enigma: revelation of What Is coupled with the subtlety of its veiling. This system of seeming contradictories was suggested in the emblem code governing garden design. To cite but one illustration:

> Water, and especially the play of water in fountains, becomes . . . the sensible measure of the vitality of the antique world. This intuition is worked out most elaborately . . . at the Villa d'Este, where water playing from the many-breasted Diana of Ephesus originally made harmonies on the famous water organ, and streams flowing from a hundred fountains connect the Fountain of Tivoli, where a sibyl pays tribute to the lord of the villa, with that of a reborn Roman Triumphans (Comito, 1978).

A major advantage the garden had over traditional oratory in its rehabilitation was the freedom it afforded the visitor to alter his vantage point at will and thus to widen it. In this, the garden shared with pageants and festivals the ability to display spectacle while the beholder's point of view shifted so as to absorb a richer array of visual impressions than words alone could offer. As the viewer wandered in the garden, he came upon symbolic tableaux from which he might serendipitously extract meaning. Additionally, walking encouraged contemplation, because it paralleled the process of rhetorical invention, the mental search in which the orator ransacked the *topoi* of his memory for ideas.

Perpetual flexibility enabled the walker to grasp more readily the symbolic meaning than would ritual or formal instruction.

> Any appropriate stone or tree trunk or flower can become without warning an appropriate space on which another order of signs supersedes nature's original significance . . . (Cool, 1981).

Whereas public debate might force the auditor to a conclusion, epideictic highlighting led the visitor toward a more personal realization. When we "come home" to Nature, we rediscover our own nature. It was thus no accident that Ficino inscribed a quintessential epideictic statement on his garden wall:

> All things are directed from the good to the good. Rejoicing in the present you must not prize wealth or desire dignity. Flee excess, flee affairs, rejoicing in the present (Comito, 1978).

Rejoicing in the present, remaining attentive to What-Is-As-It-Is—that is the epideictic quest. Celebration is, in its origins, neither hollow nor abstract. It calls the beholder to entertain in his attention what is presented to him. It invites him to remain present to what is presented.

Both rejuvenation and broadened perspective were served by a copious style. In rhetorical theory, copiousness symbolized a kind of fecund variety, a surplus overflow that suggested vitality. The garden parallel for the copious was the fruitful, the abundant. Growth and fertility metaphorically reminded the spectator of the lively inspiration needed for the world of affairs. Architectural variety contributed to the impression in the garden.

> Steps and stairways are features intensely alive and varied. There are the stately moving steps of Torlonia and the fanciful steps of Crivelli; there are the ramps of Este that flow down and of Palmieri that sweep up to the terrace. In the Vatican the steps are grave and dignified; at Spello yawning and full of laughter. At Tremezzo they echo the ripples playing across the lake (Shepherd and Jellicoe, 1966).

The most spectacular ornamental reminder of abundance was perhaps the artful use of moving water, in cascades and free-standing fountains. Examples ranged from quiet pools designed for peace and reflection, to trickling basins, to rushing streams meant to create rustling sounds amid foliage, to the cool, splashing effects of fountain sprays to the "supreme achievement" of a sparkling, roaring cascade (Shepherd and Jellicoe, 1966; Simon, 1967).

Yet rhetorical theory recognized that unlimited embellishment led to stylistic excess. Copiousness needed to be balanced with its rhetorical counterpart, decorum, the harmonic, and appropriate ordering of details, to lend a sense of proper dignity. In the garden as in the oration, this tribute to custom and good usage also reinforced the ideal of a detached perspective, for the spectator needed a certain distance in order to discover the overriding formal pattern governing the profusion of detail.

> "The soul is delighted by all copiousness and variety," Alberti says, but copiousness "without dignity" is a "dissolute confusion"; and his prescriptions of villa gardens include, along with suggestions for pavement, statues, and artificial grottoes, rules for the proper order—no random grove—in which trees should be planted. [There is] the insistence on both *copia* and *ordo*, on sensuous variety and intellectual precision (Comito, 1978).

Surrender to nature must therefore be balanced with the detachment entailed in objectivity.

> Order is discovered in the sensuous experience only as we move away from that experience: an order less of *res* than of *verba* . . . adheres more rigidly to the rhetorician's paradigm, rejecting all the seductions of narrative that might sink [one] in mere repetition (Comito, 1978).

The delightful submission to adornment's onslaught must, if the experience is truly to commemorate in the epideictic sense, culminate in introspection, the withdrawal inward in a moment of reflection by the beholder. So it was in his realization of the correspondences between the well-ordered soul and the well-ordered garden that the visitor achieved epideictic appreciation.

We may thus conclude that the Renaissance garden fulfilled the aims of traditional ceremonial discourse: It heightened the individual's appreciation of his human being even as it rehearsed him in the rhetorical *virtu* needed to meet his obligations as a participant in civic affairs. It served those ends with devices symptomatic of the age. First, it afforded its visitors that temporary repose thought to restore mental vitality, a state often held hostage to the cares of the world in the active citizen. Second, the semiotic of the "things" of nature ready at hand in the garden—water, soil, rock, tended plant, life, shade, aroma—were rendered in icons of a well-ordered, bountiful cosmos, a reminder that we can order our minds and souls as we can our property, and so increase our virtue. The icons thus displayed performed a dual function. They lent a sense of tranquility; they were also material reminders of the need for decorum, for appreciating the norms of good order. And finally, the garden demonstrated the rewards of detachment and the ability to shift one's perspective. In this regard, the garden was entertaining; one familiar with the iconic code would have his attention drawn into a state of alertness and discovery as he decoded the puzzles adorning the environment, memorials to the beauty and latent meaning abiding throughout the world in which we all make our home. These are but a few of the means by which the garden met those authentically re-creational ends traditionally assigned to epideictic celebration. And it was in this sense that the garden operated as a ceremonial vehicle to render its users more fit to profit the republic by their prudence in word and deed.

## NOTES

[1]This paper is excerpted from a larger monograph on the topic, "Central Park and the Celebration of Civic Virtue," appearing in T. W. Benson (ed.), *American Rhetoric: Context and Criticism* (Southern Illinois University Press, 1989), 221–226.

[2]Among others see J. E. Seigel. 1968. *Rhetoric and Philosophy in Renaissance Humanism*. Princeton University, Princeton; N. S. Struever. 1970. *The Language of History in the Renaissance*. Princeton University, Princeton.; C. Trinkaus. 1970. *In Our Image and Likeness*, 2 vols. Constable and Co., London.

## LITERATURE CITED

Comito, T. 1978. The idea of the garden in the Renaissance. Rutgers University, New Brunswick.
Cool, K. E. 1981. The Petrarchan landscape as palimpsest. The Journal of Medieval and Renaissance Studies 11:96.
DeNeef, A. L. 1973. Epideictic rhetoric and the renaissance lyric. The Journal of Medieval and Renaissance Studies 3:203–231

Gray, H. 1963. Renaissance humanism: The pursuit of eloquence. Journal of the History of Ideas 24:497–514

Hardison, O. B. 1971. The orator and the poet: The dilemma of humanist literature. The Journal of Medieval and Renaissance Studies 1:33–44

Kahn, V. 1983. Giovanni Pontano's rhetoric of prudence. Philosophy and Rhetoric 16:20–23.

Lee, R. W. 1967. Ut pictura poesis. Norton, New York.

Osgood, C. G. (trans.). 1956. On poetry (G. Boccaccio). Bobbs-Merrill, New York.

Pocock, J. G. A. 1975. The Machiavellian moment. Princeton University, Princeton.

Ricci, L. (trans.). 1940. The discourse (N. Machiavelli). Random House, New York.

Shepherd J. C. and G. A. Jellicoe. 1966. Italian gardens of the Renaissance. Architectural Book Publishing, New York.

Simon, K. (trans.). 1967. Renaissance and Baroque (H. Wolfflin). Cornell Univ., Ithaca, New York.

Wilson, T. 1963. The arts of rhetoric. In: O. Hardison (ed.). English literary criticism. Appleton-Century-Crofts, New York.

CHAPTER 4

# *Can You Have a Merry Christmas Without a Tree?*

---

Brian G. McDonald

Graduate Student, Psychology, Sam Houston State University

A. Jerry Bruce

Professor of Psychology, Sam Houston State University

## *INTRODUCTION*

Review of the literature in psychology relating horticulture to human functioning revealed a sparsity of research findings. The majority of the research focused on gardening projects for geriatric and handicapped populations. The results of the investigations indicated that human interaction with plants heightened self-esteem, enhanced purposeful behavior, increased creativity, aided with adjustment to new environments, and improved clients' self-expression (Inman and Duffus, 1984–1985; Hill and Relf, 1982; Isaacs, 1986; Watkins, 1982). From this research, the usefulness of plants in shaping the way people feel about themselves and about situations can be inferred.

The human connection with plants also has an intimate relationship within the situational variable of celebration. The role of horticulture in human experience has been evident since early times as an integral part of human culture as a whole. One of the central events in which horticultural elements have been found in previous periods of history has been in the act of celebration.

Celebrations comprise one of the common threads running through human society. The word "celebrate" comes from a Latin word, *celebratio*, meaning people together. One might go so far as to say that where people get together, they are apt to celebrate. The human being as a social animal has two natural tendencies, to fight and to celebrate. Among the most common depictions in earliest cave paintings of primitive peoples were pictures of celebration or of war (Burland, 1965; Kuhurt et al., 1979). These two common themes abound in the ancient Egyptian *Book of the Dead* (Budge, 1960). In the Old Testament, again these two topics were central; as an example, the book of Psalms contained many songs for the purpose of celebration (see Psalm 96).

Our central concern in this paper is the presence of horticultural elements within the context of celebration. We contend that celebration is an integral part of life and, further, that horticultural elements have become a part of the collective unconscious of people as they enter the act of celebration.

Days of celebration are often called feast days. In other words, eating is a natural part of the process. The first celebrations were probably around the carcass of an animal killed in the hunt, but the present paper's focus is not on food (some of which, of course, is horticultural), but on the environmental context in which celebration occurs.

Some examples from the Bible illustrate this focus. In Genesis (13:18), Abraham built an altar to God under the oaks of Mamre. When the people celebrated Jesus' entry into Jerusalem, "A very large crowd . . . cut branches from the trees and spread them on the road" (Matthew 21:8).

When the time to celebrate comes, people naturally utilize horticultural elements to decorate the place of celebration along with placing special meaning and significance upon these elements: a tree for Christmas, a lily for Easter, roses for Valentine's Day, a pumpkin for Halloween, etc. The paper presents the research methods and findings in relation to the notion that horticultural elements are an integral part of the celebration of some of the most popular holidays, and that life is made more enjoyable by the presence of these elements.

## OPEN-ENDED DESCRIPTIONS

Subjects were asked to imagine an indoor and an outdoor scene of a particular holiday and to write a brief description of the scene. The holidays investigated were Thanksgiving, Christmas, and Valentine's Day. The scenes were examined for the presence of horticultural items used in the descriptions, such as leaves, trees, grass, roses, etc. The subjects completed the descriptions along with providing the following demographic information about themselves: sex; age; and classification by class, nationality, religious preference, and marital status. None of these demographics systematically differentiated the subjects' performance; therefore, they were not considered in the discussions below.

### Thanksgiving

Eighty-one introductory psychology students participated in the Thanksgiving Day task. Of the subjects, 62.96% responded with at least one horticultural element in their descriptions of the scene. In the outdoor description, 61.73% of the subjects responded with at least one horticultural descriptor, and for the indoor description only 19.75% of the subjects responded with at least one horticultural cue.

### Christmas

Forty-four introductory psychology students participated in the Christmas task: 81.82% of the subjects used at least one horticultural descriptor. In the indoor scene, the horticultural elements were used 72.70% of the time; and for the outdoor scene, 50.00%.

### Valentine's Day

Forty upper-level psychology students participated in the Valentine's Day task. Of the subjects, 80% responded with at least one horticultural element in their description, 55% of the subjects responded with at least one horticultural descriptor to the indoor scene, and 60% so responded to the outdoor scene.

## Discussion

Without any prompting, a majority of subjects for each of the above holidays used horticultural items in describing scenes of celebration. There was a degree of variability in comparing indoor and outdoor scenes, which may indicate, to some degree, the preferred location for the celebration. The presence of the horticultural elements within the descriptions supported the notion that persons associate plants with these celebrations.

# WORD-ASSOCIATION TASK

To examine the association between plants and human events of celebration, a word-association test was developed. Each subject was handed a booklet and was instructed not to open the booklet until told to do so. Each booklet consisted of eight pages; the name of a holiday was printed at the top of each page. The pages were randomly arranged in each booklet. The holidays included the Fourth of July, Thanksgiving, Memorial Day, Valentines' Day, Halloween, Easter, Labor Day, and Christmas. Subjects were given instructions to write as many words as came to mind in response to the words that appeared at the top of the page. The experimenter allowed 15 seconds per page, instructing the subjects to turn to the next page at the end of the 15 seconds. Each booklet was then examined for horticultural associations to holidays.

## Results

Thirty introductory psychology students participated in this task. The results were examined in terms of percentage of subjects listing at least one horticultural association per holiday. The findings for the various holidays were: Fourth of July, 3%; Thanksgiving, 10%; Easter, 13.3%; Halloween, 26.7%; Valentine's Day, 43.3%; and Christmas, 60%.

## Discussion

A word-association task is considered an ambiguous task, and responses are controlled by what people readily bring to consciousness. As in the first research method, i.e. descriptions of holiday scenes, the above results supported the notion that horticulture serves as an important association when people are asked to recollect certain holidays. Not all holidays are equally represented within this notion. Interestingly, both research procedures provided evidence indicating Christmas as having the strongest association with horticultural elements.

So far, support has been found for the notion that the celebration of holidays and the presence of horticulture elements seem to occur together. The question still remains as to whether or not these horticultural elements play a role in the meaningfulness and enjoyment of these holidays, or rather if the absence of such elements in a given description decreases the value, meaningfulness, or enjoyment of the holiday.

## EXPERIMENTAL MANIPULATION OF DESCRIPTIONS

The third type of research method used was the experimental manipulation of descriptions of a holiday with or without a horticultural cue, submitted to subjects for their rating of meaningfulness and enjoyment. The descriptions were manipulated for content, that is, the presence or absence of a Christmas tree. From previous findings, it was predicted that descriptions including horticultural elements would be rated more favorably than descriptions without horticultural cues.

### Procedure

Since its relative strength for eliciting horticultural elements was stronger, Christmas was selected as the holiday referred to in the descriptions. Three descriptions were created with one of two conditions: (1) with the inclusion of a horticultural element, and (2) without a horticultural element. Each of the three descriptions of Christmas depicted scenes from different periods: pre-Renaissance England, nineteenth-century Europe, and present-day Christmas in America.

Subjects were randomly administered the three descriptions, one from each period, and were instructed to rate them on the two scales that appeared below each description. Each description was on a separate sheet of paper. The first scale rated meaningfulness and the second scale rated enjoyment on a seven-point scale, in which 1 was being very meaningful or enjoyable and 7 represented not meaningful or not enjoyable.

### Results

Sixty introductory psychology students participated in this task. Each subject was given three descriptions for rating. In the analysis of results, some ratings were randomly deleted from the data pool in order to produce equal $n$'s. The data were collapsed across different periods. A t-test was performed on the results from the two conditions for the meaningful measure, the enjoyment measure, and the total score. The differences between the with-tree and without-tree conditions for the three dependent measures were only marginally significant (meaningfulness, $t = 1.79$, $df = 77$, $p = .0759$; enjoyment, $t = 1.64$, $df = 77$, $p = .104$; total, $t = 1.93$, $df = 77$, $p = .0554$).

### Discussion

The results tended to support the hypothesis that horticultural elements enhanced the meaningfulness and the enjoyment of Christmas. Subjects tended to rate the descriptions that included a horticultural element as more meaningful and enjoyable than descriptions that excluded horticultural elements. The marginal nature of significance within the results may have been due in part to an unconscious projection of the Christmas tree into the verbal descriptions used, regardless of the presence or absence of the tree in the description. This conclusion is supported by the previous findings demonstrating the relatively strong ability of subjects to respond with horticultural elements to ambiguous stimuli.

To sum up the tentative findings of the research presented in this paper, one may with caution advance the proposition that, although it may be too early to say, "You cannot have a merry Christmas without a tree!", perhaps one can say, "Your Christmas will be merrier with a tree!"

## LITERATURE CITED

Budge, E. A. W. (trans.). 1960. The book of the dead. Bell Publishing, New York.

Burland, C. 1965. North American Indian mythology. Paul Hamlyn, London.

Hill, C. O. and P. D. Relf. 1982. Gardening as an outdoor activity in geriatric institutions. Activities, Adaptation and Aging 3(1):47–54.

Inman, M. and J. Duffus. 1984–1985. Adaptations to dwellings and interiors by independent older adults following relocation. Journal of Housing for the Elderly 2(4):51–61.

Isaacs, A. F. 1986. How to be personally creative: Self help to good mental health. Creative Child and Adult Quarterly 11(1–2):119.

Kuhurt, A., C. Fagg, F. M. Clapman, and E. Wiltshire. 1979. Atlas of the ancient world. Crescent Books, New York.

New York International Bible Society. 1978. The Holy Bible. New International Version.

Watkins, C. E. 1982. The counselor's house of plants. School Counselor 30(2):137–138.

CHAPTER 5

# The Role of Flowers in the Bereavement Process

Candice A. Shoemaker,* Diane Relf,† and Clifton Bryant ‡

*Research Associate, Department of Horticulture, † Associate Professor of Horticulture, ‡ Professor of Sociology, Virginia Polytechnic Institute and State University

## INTRODUCTION

An important sector of the florist's business is sympathy flowers and funeral tributes. Flowers have been an integral part of funeral ceremonies for centuries. Although they are still a component of the funeral, florists are seeing a decline in their sympathy sales.

Trends in funeral traditions are changing. More people are foregoing the traditional funeral and thus the accompanying rituals. A few decades ago, it would have been unthinkable for close family, friends, and even neighbors not to send funeral flowers. That attitude seems to have changed. Various factors have been postulated that may contribute to this change: for example, the changing pattern of social and family life, the smaller size of families, the geographic dispersal of family members, the decline in the neighborhood feeling, the decline in religious belief and church going, the increase in cremations over burials, and the increased use of the request for "memorial donations in lieu of flowers" (Indepth Research, 1984).

The importance of the ritual of death and bereavement seems to have lessened. Although professionals in the field of death education and counseling agree that grieving is essential in adjusting to the loss of a loved one, there remain numerous questions about how best to facilitate successful grief work. Traditionally, ritual has been an important element of the bereavement process, although very little information is known on how or even whether the various funeral rituals facilitate grief work. Wilcox and Sutton (1977) report that rituals help make sense of death and lend reality to the loss of a loved one. Post-funeral rituals are also important in the grieving process. Bolton and Camp (1989) evaluated the association of post-funeral ritual acts and adjustment and reported that there was a link between specific ritual acts and higher adjustment scores.

The questions we are asking are: Do flowers serve a role in the bereavement process, especially in the context of this changing background of behavior and attitudes towards death and funerals? If they do serve a role, what is this role or roles?

At this writing, this research project has not been completed. The material presented here focuses on the preliminary work and methodologies used for this project. The data for this study will be collected by a questionnaire mailed nationwide to bereaved persons. Prior to mailing the questionnaire, a good deal of preliminary work was done to orient ourselves in an area where very little information is available and to determine the most effective way to collect this type of data. The preliminary steps included surveys of funeral directors and grief therapists, personal interviews with psychologists and sociologists, and focus group interviews with individuals who had experienced the loss of a loved one in the past five years.

## SURVEY OF FUNERAL DIRECTORS

A questionnaire was distributed at the National Funeral Directors Association annual convention in October 1989. A booth was set up in the registration area of the convention center, which was in the lobby to the trade show exhibit. Individuals could complete the survey at the booth or take it with them, and return it to the booth before the close of the convention. The majority completed it at the booth. Many were also willing to be informally interviewed by the researcher after completing the questionnaire. The researcher was at the booth during the exhibit hours to answer questions and conduct interviews. Surveys were available at the booth at all times during the convention.

The results from this survey suggest that flowers serve two very different roles: an emotional role and a functional or utilitarian role. First, regarding the emotional role, the funeral directors felt that flowers provided comfort to both the sender and the receiver. The ritual of sending flowers provides a way for the sender to express sympathy and care to the bereaved. The funeral directors also thought that flowers provided comfort and warmth to the bereaved during the visitation and funeral service. Fifty-three percent of the funeral directors said that the bereaved mention flowers as a comforting aspect of the funeral immediately after the funeral. Eighty-three percent believe that months or even years after the funeral, the bereaved recall flowers and plants as a comforting part of the funeral.

The second role that flowers have in the bereavement process that was suggested from the funeral directors survey was a more functional role, that is the flowers are noticed in very tangible ways. All of the funeral directors surveyed said that flowers are generally discussed, looked at, touched, smelled, or talked about during the visitation or funeral service. Some of the comments most often heard about the flowers are how many flowers there are, the design and color of the arrangements, and the condition of the flowers. Seventy-four percent of the funeral directors said that people want to know the names of the flowers and plants during the visitation or funeral service. As well as noticing the flowers at the funeral, all the funeral directors said that family members took some of the flowers and plants home with them after the funeral service.

The funeral directors were surveyed about the role of flowers and plants in the funeral ritual. Therefore, they were asked questions not only about flowers but also about the effectiveness of the other rituals associated with a funeral, such as visitation, a traditional funeral service, memorial contributions, and sympathy cards. The results indicate that flowers are an important component of the funeral ritual, but how important are they in relation to all the other rituals? To answer this question, a questionnaire was developed for grief therapists.

## SURVEY OF GRIEF THERAPISTS

A random sample of 400 grief therapists associated with hospice were provided by the National Hospice Association. The questionnaire consisted of a list of 11 items associated with the funeral ritual. The grief therapist was asked, on a scale of 1 to 6 with 1 being not at all and 6 being a great deal, how much each of the items helps in the grieving process. There were also three open-ended questions regarding the use of flowers in the funeral. The results from this survey will indicate how effective flowers are in helping in the grieving process compared to other elements of the funeral ritual and bereavement process. The responses to this survey have not been analyzed yet.

## FOCUS GROUP INTERVIEWS

The information learned from the funeral directors survey and interviews and personal interviews with psychologists and sociologists that conduct research in this field was used to develop a questioning route and interview protocol for focus group interviews. A focus group is a carefully planned discussion designed to obtain perceptions on a defined area of interest in a permissive, nonthreatening environment (Krueger, 1988).

There were three objectives for these focus groups. First, to acquire information to build a survey instrument. Second, to collect information on how sympathy flowers aid in the grieving process. Third, to collect information on what sympathy flowers mean to the bereaved.

Four groups were interviewed. Participants in two groups represented the community at large, and participants in the other two groups were students from the introductory psychology classes on campus. The participants had to have experienced the death of a loved one in the past five years, the death having been no more recent than one year ago.

All of the sessions were conducted on the campus of Virginia Polytechnic Institute and State University. Two interviewers, a primary interviewer and support interviewer, were at each session. Due to the sensitivity of the subject being discussed, a grief therapist was present at each session.

The questioning route and interview protocol consisted of five broadly stated topic questions, as well as preplanned probes for each major topic. The same questioning route was used for each group, but it was only used as a guide. The format of a successful focus group differs from that of the traditional question-response interview in that a lot of discussion takes place between the members of the group and the interviewers function more as moderators. Therefore, we allowed the personality of each group to develop, but the interviewers made sure that the group focused on the five topic areas we had developed. We did cover all five topics in each group so that we could make comparisons across groups in the analysis phase.

All of the sessions were audio-taped and transcripts were made. The transcripts are the basic data that this type of research produces. There are two basic approaches to analyzing focus group data: first, a strictly qualitative or ethnographic summary and, second, a systematic coding via content analysis. Both approaches were used for this project.

Both types of analysis were performed by three people; the two interviewers and someone who had not been involved with the project prior to reviewing the transcripts; the nonbiased third person provided internal validity to the analysis.

The five topics from the interview protocol were used for organizing the topic-by-topic analysis of the discussions. All three of us read each of the transcripts and independently rated each statement based on the five topics from the interview protocol. The statements were then sorted based on these five topic areas. The next step is to code categories within each topic area and locate items that can be systematically counted. The systematic approach used in analyzing the results from the focus groups will strengthen the instrument for the nationwide survey.

Although the analysis is not complete, it appears that the themes brought out from the funeral directors survey will be supported by the results of the focus groups. That is, flowers are a comfort during the visitation and funeral service and they also serve a very functional role, mainly giving people something to talk about or focus on other than the deceased.

## NATIONWIDE MAILED SURVEY

A random sample of 5000 names and addresses were selected from a data base of 120,000 widows and widowers. The data base consisted of individuals who had responded to a lifestyle change questionnaire in 1989.

The 35-item questionnaire consists of 4 sections. The first section focuses on many of the rituals associated with death and funerals and how effective they are as an aid in working through grief. The next two sections contain questions about the use of flowers in the funeral. The last section contains demographic questions.

The results from this questionnaire will provide information on the importance of sympathy flowers in relation to many of the other rituals surrounding death and funerals, as well as clarifying the role of sympathy flowers in the bereavement process.

## LITERATURE CITED

Bolton, C. and D. J. Camp. 1989. The post-funeral ritual in bereavement counseling and grief work. Journal of Geriatric Social Work 13(3/4):49–59.

Indepth Research. 1984. Report on a small-scale qualitative research study to explore, examine and assess consumer attitudes, feelings, needs and motivations towards sympathy flowers. REF: GAE/357, London.

Krueger, R. A. 1988. Focus groups: A practical guide for applied research. SAGE Publications, Los Angeles.

Wilcox, S. G. and M. Sutton. 1977. Understanding death and dying: An interdisciplinary approach. Alfred, New York.

CHAPTER 6

# *Vita Brevis:*
# *Moral Symbolism from Nature*

Joseph C. Cremone, Jr.

Clinical Instructor in Surgery, Harvard Medical School

Richard P. Doherty

Horticulturist, University of Massachusetts/Boston

Botanical paintings were often intended to awaken moral consciousness. Ethical and religious messages were conveyed through symbolism associated with flowers. One flower might serve as a reminder of the transience of life; another could symbolize the eucharistic doctrine of Resurrection. Moral lessons against pride and greed blossomed in a painting and were intended to take root in the viewer's mind. Have such floral paintings lost the impact of their moral statements now that many of the emblematic meanings of flowers have largely been forgotten? Do the lessons these paintings convey have any social relevance today? Finally, are similar moral and social messages expressed in modern art through botanical symbolism? These questions are intimately related and form the basis of this investigation into floral symbolism in painting.

The Renaissance art historian Vasari acknowledged such botanical symbolism when he commented on the figure of Venus in Botticelli's *Primavera*: "Venus," he observed, "whom the Graces are covering with flowers, was a symbol of Spring." The *Primavera* contains some five hundred individual plants, flowers, and grasses. Their allegorical significance is extremely varied and, at times, might seem incongruous. Some flowers, for instance, symbolize death whereas others refer to love and marriage. Why such a bouquet of seemingly disparate symbolism?

It has been suggested that the *Primavera* was completed for Lorenzo de Medici before his marriage and after the assassination of his brother Giuliano (Levi d'Ancona, 1983). Botticelli therefore combined the buttercup, an emblem of death, with other flowers that were associated with love and marriage: the cornflower, a symbol of a beloved woman; the violet, symbolic of love bestowed by Venus; and the pink, traditionally included in wedding bouquets. Under the feet of the central Grace is found borage, which brings happiness in love, and also fennel, another emblem of love. Botticelli added jasmine, the symbol of elegance and grace, and the flowers of the wild strawberry, which denoted seduction and sensuality.

The figure of Flora, a personification of marriage, is robed in a bouquet of pinks, cornflowers, roses, and daisies, all flowers symbolic of marriage. According to Greek legend, the cut roses in Flora's arms were created at the birth of Venus; they therefore make reference to sexuality. Three types of love were represented in the *Primavera*—carnal, human, and divine—and each was emphasized through floral symbolism. And Botticelli's spiritual message? Mercury looks away from the three Graces to the highest, ideal sphere of love; so too, the viewer is invited to rise by degrees to the heights of divine love (Baldini, 1986).

Just as Botticelli found symbolic beauty in the flowers of Tuscany, 17th-century Dutch artists delighted in representing the large numbers of flowers recently imported into Europe. These importations were frequently used because their vivid colors were not common or even existent in native plants (MacDougall, 1989). The striking size and shape of flowers such as the crown imperial often served as brilliant finials in pyramidal compositions. An abundance of flowers became a metaphor for the wisdom and blessings of God as found in the variety of the natural world. Flower paintings of one species alone were therefore extremely rare.

An atmosphere of transience pervades much of 17th-century Dutch still-life painting. Whereas real flowers wither and die, painted images continue to delight with *tromp l'oeil* beauty: *Ars longa, vita brevis*. These painted flowers were reminders of the transitory nature of life and the ephemeral aspects of beauty (Wheelock, 1989). A message of hope was also included in references to Christian doctrine and the promise of resurrection and eternal life. Some religious lessons were subtly implied, whereas other gospels were as explicit as the crisp rendering of each fruit and flower.

Such images of flowers as symbols of life's transience have biblical sources (Ember, 1989): "Man that is born of woman is of few days and full of trouble. He cometh forth like a flower, and is cut down; he fleeth also as a shadow, and continueth not" (Job 14:1–2). Another reference reinforces the visual metaphor: "As for man, his days are grass; as a flower of the field, so he flourisheth. For the wind passeth over it, and it is gone . . ." (Psalm 103:15–16).

Dutch floral still-life paintings were thus didactic expressions of ideas concerning death and transience. These floral paintings are known as *vanitas* still-lifes, the genre taking its name from the line from Ecclesiastes 1:2, " 'Vanity of vanities,' saith the Preacher, 'all is vanity.'" In some paintings, a Bible was opened to Ecclesiastes to emphasize further the moral lesson. A human skull might be included as a symbol of the transience of life. A rose placed near the skull reminded the viewer that human life, like that of the rose, is short. Other *vanitas* images were included to develop and highlight moral and spiritual lessons. A fly crawling on a fruit, for instance, was a reference to death, as were the insects shown eating the flowers and leaves. Bird's eggs, in contrast, were often represented to suggest the continuity of life.

An extensive emblematic vocabulary gave abstract associations to flowers and enabled artists to develop numerous religious, social, and economic references. This "disguised symbolism" was explained in contemporary emblem books—books of symbolic pictures with explanatory texts. One example is a 1614 emblem depicting two large tulips of the rare, striped variety. The accompanying inscription gives a clue to the symbolic meaning of tulips in Dutch floral pieces. It reads: "A fool and his money are soon parted" (Foshay, 1984). The inclusion of tulips in still-life paintings was thus a warning of the consequences and potential financial losses associated with greed and speculation. Highly prized red-and-white tulips were often added to the arrangements as references to the tulipomania of the 1630s that led to the bulb market collapse and subsequent economic recession. The record-breaking prices of the contemporary art market recall the bulb speculation of the 17th century. Living bulbs were then coveted; now painted flowers cause the gavel to crack at Sotheby's.

The art of Andy Warhol might seem far removed both in period and intent from the *vanitas* compositions of the 17th century. Warhol's *Flowers* were not derived from a scientific study of nature, as were the Dutch paintings. He took his floral image from a photograph of a hibiscus in a June 1964 issue of *Modern Photography*. *Flowers* presents a phosphorescent beauty that gradually fades and turns tragic. The response is similar to that elicited by the flies, snakes, and withering blossoms seen on close inspection in Dutch *vanitas* paintings. The contrast between the garish colors of the flowers and the blackness of the background contributes to the ultimate sense of despair and death of Warhol's *Flowers*. "The garish and brilliantly colored flowers," wrote one critic, "always gravitate toward the surrounding blackness and finally end up in a sea of morbidity. No matter how much one wishes these flowers to remain beautiful they perish under one's gaze, as if haunted by death" (Coplans, n.d.).

The sinister beauty of Warhol's *Flowers* relates them to his *Disaster Series*, silkscreens of suicides, car crashes, and electric chairs, all haunting comments on contemporary society. Warhol presented his *Flowers* to a society that has polluted its environment and knows nature largely through mass-produced printed images. Warhol's *Flowers* have thus become the Pop Art *vanitas* images of our own times. Flowers continue to convey a social message, but the lesson is now written in Day-Glo colors.

Flower imagery has provided a boundless wealth of symbols enabling artists to convey diverse attitudes and philosophies. Such associations have applied veneers of metaphorical meaning to paintings of flowers and to the flowers themselves. The symbolic meanings frequently mutate in different periods and cultures. A garland of roses around the neck of Venus, for instance, invites references to love and sexuality. The same bouquet offered to the Madonna and Child represents Christian symbols of pure love and sacrifice. It is the beauty of the flowers themselves, however, that ultimately elicits the greatest aesthetic response. Placed before a floral painting with emblem book in hand and eager to decipher hidden symbolism, the viewer should recall the tempering words of La Fontaine: "Beware as long as you live of judging by appearances."

## LITERATURE CITED

Baldini, U. 1986. *Primavera*: The restoration of Botticelli's masterpiece. Harry Abrams, Inc., New York.

Coplans, J. n.d. Andy Warhol. New York Graphic Society Ltd., New York.

Ember, I. 1989. Delights for the senses: Dutch and Flemish still life paintings from Budapest. Budapest Museum of Fine Arts, Budapest.

Foshay, E. M. 1984. Reflections of nature: Flowers in American art. Alfred A. Knopf, New York.

Levi d'Ancona, M. 1983. Botticelli's *Primavera*: A botanical interpretation including astrology, alchemy and the Medici. L. S. Olschki, Florence.

MacDougall, E. B. 1989. Flower importation and Dutch flower paintings, 1600–1750. In: Arthur K. Wheelock, Jr. (ed.). Still lifes of the golden age. National Gallery of Art, Washington, DC.

Wheelock, A. K., Jr. 1989. Still life: Its visual appeal and theoretical status in the seventeenth century. In Arthur K. Wheelock, Jr. (ed.). Still lifes of the golden age. National Gallery of Art, Washington, DC.

CHAPTER 7 – ABSTRACTS

# Educational/Sustainable Environmental Gardens in a Man-Made Agrarian Landscape

John C. Billing

Professor of Park Administration and Landscape Architecture,
Texas Tech University

## ABSTRACT

Prior to the settlement of the Old Northwest Territory, the landscape was blanketed with a great, enclosed forest canopy interrupted by a series of clearly-defined open spaces. The openness of these isolated, treeless spaces were in sharp contrast to the enclosed, special feeling provided by the forest. To the new settlers the diversity of plants and the spaces they created were of little importance. Cleared, open land harbored wealth and capital; the forest and its environs was nothing more than another obstacle to overcome on the way to a better life.

This presentation explores the relationship between early 1900s, historical field-mapped, plant community data of those isolated, open spaces and specific soils data developed in the 1930s. Understanding the correlation between these two factors would provide the opportunity for current, man-made agrarian landscapes to become much more diverse and regionally significant. These spaces have the opportunity to become a series of self-sustaining, environmental/educational gardens in association with the open agrarian landscape of the Midwest.

# Home Gardens in Honduras

Lynn Ellen Doxon

Assistant Professor of Horticulture, New Mexico State University

## ABSTRACT

Many projects promoting home gardening have been done in Honduras, but only a few have had lasting results. I was invited to Honduras to examine why these home garden projects were unsuccessful. Previous projects were approached with the assumption that Hondurans do not have home gardens. Nonetheless, I found intensively cultivated home gardens around almost every house. Seventy-five food species were produced in these home gardens. They were not generally recognized as home gardens by development workers because the food species were dominated by ornamental species. Ornamentals were found to have significant social importance in Honduras. For many Hondurans, these flowers were the only attempt to beautify their surroundings. Recommendations were to promote improvements in varieties and culture of the plants already there and to include ornamentals in the garden plans rather than trying to use standards from another place and climate to develop project plans.

# SECTION II

# *PLANTS AND THE COMMUNITY*

CHAPTER 8

# Effects of Plants and Gardening in Creating Interpersonal and Community Well-Being

Charles A. Lewis

Research Fellow in Horticulture, Morton Arboretum, Lisle, Illinois

## INTRODUCTION

In preparing for this presentation, I decided to learn more about community development, a process by which a community gains a sense of itself, looks at its problems, and decides to take steps to correct them. For years, I have reported on changes in people and neighborhoods as a result of gardening (Lewis, 1973, 1980), but did not know how they fit into the larger concept of community development.

The term *community* refers to people who live in some spatial relationship to one another and who share interests and values (Carey, 1970). The community might be a neighborhood, housing project, school, prison, or other spatially defined relationship. The collective attitudes of its members endow each community with a characteristic life and personality of its own, ranging from vibrant to lethargic.

The total physical condition of a community, its buildings, vacant spaces, and streets, makes an enormous difference in how members of that community feel about themselves. What we see often tells us what we are.

> Walking by and through trash-filled empty lots and vandalized school yards, walking along littered streets with no trees, seeing only pavement and brick when looking out the window makes people feel bad about where they live and about themselves. They know they do not matter, that their pleasure or comfort is unimportant (Primak, 1987).

Lyndon Johnson said,

> Ugliness can demean the people who live among it. What a citizen sees everyday is his America: If it is attractive, it adds to the quality of his life, if it is ugly it can demean his existence (Johnson, 1965).

The physical condition of a community, therefore, plays a double role; for the community, it is a measure of itself; for outsiders driving through the community, its physical condition creates an impression of its quality and character.

Ideally, a community should have a clear sense of itself and seek cooperatively to improve its physical, economic, and social conditions. Frequently, residents of low-income neighborhoods are not acquainted with each other, feel isolated, and therefore feel a sense of powerlessness. A community activity such as gardening can be used to break the isolation, creating a sense of neighborliness among residents. Until this happens, there is no community, but rather separate people who happen to live in the same place.

A project such as gardening is usually initiated by a charismatic leader who attracts the residents. The project produces not only visible physical changes in the community but also an inward change in the residents as well. More important than the project is what happens to the people.

The initial goal is to create a place where neighbors can garden. Establishing the garden requires a leader who organizes and inspires the neighborhood to believe that together they really can create the garden. The achievement of establishing and operating that garden leads residents to realize they can achieve a change, and that perhaps there are other aspects of the community that they might also improve.

An example of gardening used as a technique for community development is found on Chicago's West Side in East Garfield Park, a decaying ghetto. Here, in a 40-block area, is Fifth City, a community development project of The Institute of Cultural Affairs, whose objective is revitalization of the community by those who live there. ICA seeks to stabilize the area by encouraging residents to stay, rebuild the neighborhoods, and create an economic base through locally owned enterprises. Gardening is used as an initial technique to encourage people to work together. Mark Welch, a leader in this project, emphasized the need for creating fast, highly visible changes as rallying points to maintain interest until slower more substantial changes can be made. If people do not know each other, they cannot start a successful community program. If people do not draw together, the building or block will not be safe. He finds that gardening, cleaning up, and creating play lots are highly visible work projects that help to unify a block: "Do something that people can see."

An example of his experience is the 3500 block of Van Buren, which came together under the leadership of a resident, Lee Haley, a retired metal worker. After he and his neighbors had invested their time and money gardening together, they started patching the sidewalk, planted 25 street trees, and finally made applications for housing rehabilitation loans for nine houses on that block. Mr. Welch says that gardening is a technique for bringing people together in a viable way. It is the glue that holds a block together until long-term economic and social development can take place. Because it draws people together and helps them gain confidence in their ability to make changes in their surroundings, gardening is a powerful tool for coalescing a sense of community.

## PROCESS OF GARDENING

Gardening is a process. Its products—plants, flowers, lawns, shrubs—are easily seen, but what do we know of the process that produces them? The process of gardening includes all the thoughts, actions, and responses from the time the gardening activity is first contemplated, through the planting and growth of the seed, to the mature plant. Personal feelings and benefits can be seen as by-products, effects unintentionally produced by the process. To discover the human factors, we shift our focus from the plants and flowers to look more closely at the person who grows them.

At each stage, the process presents the gardener with opportunities for subjective personal involvement. Something of the human spirit is invested in the gardening process. Deciding to grow a plant, sow a seed, or design a garden creates mental expectation of what the garden will look like. Because of the time required for plants to grow, the gardener must wait for tangible rewards until the garden finally blooms or bears fruit. The ultimate quality of the person-plant experience is dependent on the degree of personal involvement and success of the horticultural project. The feelings experienced during this process—hopes, frustrations, joys—are vital human components.

How might plants enter into human experience? First, we might be *observers*, seeing plants in our surroundings as they appear in parks, gardens, along streets, or in vacant lots. Though providing purely a visual experience, plants often evoke subjective responses in the viewer. Do we like what we see, is it exciting, peaceful, interesting, pleasing, boring? In another mode of experiencing plants, we might be *participants*, intimately involved with the plants being grown and directly responsible for the well-being of the plants.

The distinction between observer and participant is not precise. Consider how a camera "sees" a setting and how a person might see the same setting. We do not see as a camera does, capturing the image on film. Two cameras fitted with the same lens and film, placed in the same spot, will produce identical pictures; however, the same tree or setting may produce different responses in each person who views it, enticing to one person and threatening to another. Such was the case at the Morton Arboretum, where a given patch of woods that is inviting for me, indeed proved threatening to a group of visiting inner-city children. They asked fearfully whether there were snakes, tigers, and lions in the woods (Lewis, 1975).

Seeing and interpreting are instantaneous, so automatic that we do not even think about the processes as they happen. As a result, we assume that other people will see the same thing in the same way, but this is not true. The interpretation of what we see is formed against our background of knowledge and experience. Look at a chess board. If you do not know the game you will see a pattern of different colors and pieces. If you do know the game you will see strategies and possibilities of different moves.

Throughout life, each of us accumulates and stores personal interpretations of what we see. The brain calls on these interpretations to aid in analyzing images that it receives from the optic nerve and retina, which are considered part of the brain (Gregory, 1966). In an instant, the brain decides on the meaning of what is seen and immediately transmits chemical and electrical messengers throughout the body that trigger affective responses—feelings of pleasure, displeasure, awe, fear—and their appropriate physiological responses, such as the release of adrenaline in threatening situations. Merely looking at a tree or flower initiates this chain of visual and mental processes. Through this evaluation process, the observer is also participating, mentally if not physically.

Beyond visual observation, there is also physical participation, as a gardener, growing and caring for plants. This physical, participatory involvement creates a more intimate nurturing relationship between person and plant, which strikes a deeply personal chord. If the plant wilts, the grower waters; if it lacks vigor, the grower adds fertilizer or provides additional light, and so on through the many nurturing activities. By close observation, the grower learns to understand plant responses as a kind of language by which the plant signals its needs. A successful gardener is also a keen observer whose brain calls on past experiences to help in making decisions—that's how the thumb turns green.

The extended encounter between person and plant often leads to subjective personal feelings. When the plant grows, the gardener feels successful and proud. If it does not grow, the gardener feels sadness or perhaps even anger. Plants, whether we are growing them or seeing them, call forth a wide range of human responses.

From a human perspective, the strength of gardening lies in nurturing. Caring for another living entity is a basic quality of being human. The deep personal feelings engendered by nurturing are recorded in one's brain as part of the gardening experience. There are probably many aspects of the process of gardening that resonate profoundly in one's being and are the basis for the ability of gardening to create human well-being. These feelings, engendered in people by plants, are often so subtle as not to be apparent. We must look beyond the trees, shrubs, flowers, and vegetables and transcend the physical to become aware of personal kinds of meanings in people-plant interactions. We look now at how these personal feelings are expressed in a range of settings.

## URBAN LOW-INCOME HOUSING

Just as the light of a candle can be seen more clearly in a darkened room, so can the human benefits of plants be seen more easily in communities lacking in economic and social opportunity. Gardening projects in impoverished urban neighborhoods have produced a continuing history of amelioration and healing in those communities.

A most obvious example is found in the procession of high-rise public housing units found in central cities. In its original intent, public housing was to offer a first step toward independence, as temporary residence for families who were moving upward toward economic security. Public housing soon became, however, a permanent kind of residency for families and individuals who could not escape from poverty.

Much of public housing has not worked. Indeed, it has been the stage for expression of great social problems. Public housing does not provide residents with a sense of home and security. Crime rates in public housing are greatest in high rise towers, where public access extends from the grounds outside, through the lobby to elevators, stairways, and corridors, finally stopping at the apartment door (Newman, 1972). Residents must proceed through these desolate public areas before arriving at the safety of home, a locked apartment door.

With buildings so large, occupied by many people, there is little chance for residents to come to know each other and thereby be able to recognize strangers as intruders in the building. There is little opportunity for neighbor surveillance, of people coming and going—such as would be possible in a private home. A criminal may enter these buildings and not be challenged as a stranger.

The high tower forces a dislocation between life inside the apartment and activities on the grounds below. Mothers cannot look out the kitchen windows to oversee their children playing in the yard. The open grounds, designed as amenities, become a no-man's-land of gangs and drug dealing, where residents are afraid to enter. The anonymous open grounds provide no sense of territoriality or indication of private space. Only fences or visible evidence of individual activity, such as gardens, can create the needed delineation.

Yet in these residences across the nation, flower and vegetable gardens help to correct some of the inherent problems. In 1962, the New York City Housing Authority, largest landlord in the world, was one of the first to initiate a tenant gardening competition. It is simple in concept and operation. Residents of a building who wish to garden apply to the Authority to obtain a plot at their building and funds for purchase of seeds, plants, and fertilizer. The Authority digs up the beds and from then on the residents are on their own, planning, planting, and caring for the gardens. Gardening expertise is available from cooperative extension and horticultural groups in the city.

In August, judging teams of horticulturists, garden writers, and landscape architects select winning gardens at each housing project site. The whole program culminates in an awards ceremony at which slides of the winning gardens are shown and a representative of each garden receives a handsome award. All participants receive certificates of recognition for their participation. Excitement at the awards ceremony runs high, and everyone leaves the meeting in high spirits, ready to enter the contest the following year. This program, initiated in 1962, continues today.

Flower and vegetable gardens in public housing are easily seen, but to discover the human meanings of these gardens one must talk with the gardeners. As a judge, I have often spoken with contestants at their gardens and thus gained an insight into what the gardens mean to the gardeners. One of the first things I learned was the intensely personal feeling each participant developed towards the group's creation. At a pleasant garden in lower Manhattan, a meek woman with a Spanish accent said proudly, "They told me you couldn't grow flowers on Avenue D, but I wanted to try. Now you should see how the old folks come out every day to sit and enjoy the flowers." Another constant told me, "This is the first creative thing I have done in my life," adding that she had gone to the library to read every gardening book available. An elderly lady with a group of youngsters had planted a garden which included tuberous begonias and cotton, plants not usually seen in the city. She labeled the plants so that the children would "know the names of all the flowers, even the ones that do not grow up here," she said, referring to the cotton. This woman from the South was trying to share some of her cultural heritage with the children.

Gardeners would reveal pride in their accomplishment in several ways. When judging the gardens, it was not unusual to suddenly find someone at my side who would urgently whisper, "Isn't this the best garden you have ever seen? Don't you think we should get first prize?" The same pride would be evident in the faces at windows above the judges, watching our every move to make certain that we fully appreciated the wonder they had created. In letters to the Authority, tenants say, "what is more important is everyone getting to know each other; everyone smiles and discusses our garden; there is too much rain; not enough rain; we are all so pleased that the children are interested in caring, not destroying. From early morning till late at night you can see neighbors leaning over the garden fence. It has become the center spot of our court where everyone is a friend."

I wondered why, in the inner-city, where vandalism is extreme, fragile gardens could grow seemingly unmolested. The answer came from the gardeners. One group said, "We know who the troublemakers are, so we invited them to join our group and assigned them the job of guarding the garden—now we have no more problems." Others told of residents joining together to patrol the garden, mothers sitting with their babies near the plot, boys and girls taking turns watching. Tenants in tall buildings are assigned times to watch the gardens, keeping the gardens under constant surveillance, all ready to sound the alarm if anyone tries to destroy the plants. It was all summed up by a woman on the East Side who said that she expected no vandalism because, "All the rotten kids are in the contest this year!" No one asked for police protection. Residents, understanding the dynamics of their neighborhood, learn how to protect what they consider important through cooperation with each other.

After a few years, the program started to produce unexpected results. Public housing tenants who grew flowers and vegetables joined together in a warm kind of pride and neighborliness. They saw to it that gardens and surrounding areas were kept clean and neat. They urged the groundsmen to mow lawns adjacent to gardens more frequently. Building managers reported that at buildings with gardens, children who usually trampled grass were cultivating and watering, tenants no longer threw bags of trash or garbage out the windows ("airmailing the garbage"). Vandalism was reduced outside and inside the buildings.

Where it had been common experience to see new landscape plantings around public housing destroyed, tenants began to ask the Housing Authority for permission to help in landscaping the buildings. Inspired by their summer contest, tenants in several projects contributed their own funds for the planting of spring flowers, tulips, and daffodils, which would bloom before the contest gardens are planted. Tenants asked permission to install planters, which they would maintain, in the lobbies of their buildings. They even organized garden clubs in many of the projects. Thus, it seemed that the experience of gardening helped residents achieve a more positive attitude about themselves, their buildings and grounds.

The Chicago Housing Authority initiated its flower and vegetable contest in 1974. Again, there are gardens at high-rise dwellings and again individual pride in accomplishment. The clean-up and paint-up took a new direction at Robert Taylor Homes in Chicago. A red and white garden inspired tenants of the impersonal high rise buildings to paint entrance pillars, benches, and chains bordering walkways in matching colors. Down the street, pillars were painted blue with white stars to match a blue and white garden. Soon large murals, geometric or pictorial in design, appeared on the walls of other buildings at Robert Taylor. The anonymous decorations (CHA rules forbids painting of buildings) were exceedingly well designed and carefully executed. During the ensuing year, no graffiti appeared on the decorated portions of the buildings.

Philadelphia has long been a center for the use of gardening in community development activities. As early as 1953, Louise Bush-Brown initiated a window-box program under the guidance of the Neighborhood Garden Association, which clearly demonstrated the power of gardening to change people's attitudes about themselves and where they live. For any block in which 85% of the residents agreed to plant and maintain window boxes for two years, the Association provided contact with a garden club which brought plants to help start the window boxes. Soon colorful flowers highlighted these inner-city streets. Also quite soon, neighbors banded together to clean up the streets, and whitewash curbs, front steps, and windowsills. They converted vacant, debris-laden lots into playgrounds and gardens. These activities were spontaneous, not part of the program, but somehow inspired by the window boxes.

Why should planting petunias in a window box lead to cleaning up streets?

Part of the answer lies in the new spirit born in the gardens. It is reflected in the gardener's comments: "Before it was just a house, now it looks like home." "I've lived on this block for fifteen years. Its so nice to come to know the names as well as the faces of the other people on the block. I never knew them before." One person who lived next to what had been a debris-laden lot commented: "This had been the most dumpified place I had ever seen. Now it even smells good." "I guess I'll wash my windows now" (Bush-Brown, 1969).

The early work of the Neighborhood Garden Association continues today in over 100 blocks as part of the Pennsylvania Horticultural Society's Philadelphia Green: a community self-help program bringing plants and people together, which includes a range of activities, garden blocks, street trees, vegetable gardens, sitting gardens. On a larger, more comprehensive scale, their Greene Countrie Towne program reaches out to larger neighborhoods and city areas. The introduction to their manual reveals a distinct community development program based on gardening.

> A concentration of community gardening activities in a neighborhood inspires significant changes in its appearance, strengthens community organizations, engenders pride, and often empowers other community development activities. . . . While each neighborhood is unique in its physical features and presence, many are plagued with trash-strewn vacant lots and streets of macadam and cement that underscore the harsh urban environment. Transforming these vacant lots and barren streets into gardens and green sidewalks creates more hospitable human spaces. Residents on green blocks are more likely to clean up their streets, paint and improve their homes. The city government benefits as residents work with municipal agencies to keep the neighborhood cleaner. . . . The projects have also produced neighborhood leadership that supplements existing political and organizational leadership—gardeners who coordinate greening projects and who encourage their neighborhoods to take up the trowel. This gardening leadership is the most important factor in the ongoing success of neighborhood greening (Bonham, 1988).

In Boston, the Boston Urban Gardeners, BUG, have developed an extensive, multifaceted program to support the many gardening initiatives in that city, ranging from neighborhood gardens to large community plots, school programs, education programs, and many more. Within the city limits, there are over 120 community gardens, ranging in size from 1000 square feet on small residential streets to 10 acres on state-owned agricultural land. Together, Boston's urban gardeners grow approximately $1 million worth of fresh produce each year. These programs benefit their participants in more than one way. In *The Greening of Boston*, Primak (1987) notes,

> Beyond these practical benefits is a psychological enrichment for people who grew up gardening and farming, often in another country, and moved to the city. Reaching into the soil triggers old skills and a sense of cultural roots, grounding in an otherwise alien place.

Again, comments of the gardeners reveal the deep kinds of benefits they find in gardening.

> I just love gardening because I was born doing it. I was born in the country. I love the smell of dirt, the smell of grass. That keeps me going . . . I kneel down in that dirt, take a handful of dirt and I say, this is nature. This something that cannot be duplicated. This is God's thing (Victor Pomare, Linwood Community Garden, Roxbury).

> Public Housing is a mass of glass. We started cleaning up the yard, planted flowers with kids. Now it's to the point where parents come out. When they did the garden, they didn't destroy it. They take pride in doing it. When you have your own self-accomplishment, there's no greater gift you can give yourself (Edna Boyle, Charleston Public Housing Project).

> I like the idea of being able to mix with the neighbors, and it gives me an incentive to come out. It really helps me to save money and at the same time I know what I'm eating is really fresh (Older woman, South Cove Community Garden)

> Being out here in this open space—we've got plenty of trees, a bluejay, cardinal fly by, and right around the corner, a busy street with busses, trucks—you feel, what a relief. This is a little island in this madness (Rick D'Angelis, Nightingale Commmunity Garden, Dorchester).

> Living in the city we need to be able to walk out of doors, and be able to find peace and quiet, serenity—country in the city. Whether we live in the city or suburbs, we need that piece of green space to relate to inner self; just to get back to the roots, where everything is simple and uncomplicated. And we all need that to keep our sanity (Ethel Sheppard, Roxbury).

Gardening programs exist in all our major cities. These eloquent comments, though made in Boston, can easily represent the sentiments of a vast community of Americans across the country who have found special values in the flowers and vegetables they grow.

## WHY DOES IT WORK?

How are we to understand the cleaned streets, reduced vandalism, painted houses, new neighborliness? How did gardening interact with human spirit to produce these results? What are the people-plant factors at work?

The strong rehabilitative effect of self-esteem, particularly in a demeaning environment, was pointed out by Dr. Edward Stainbrook, Department of Human Behavior, University of Southern California School of Medicine, who said,

> An environment of ugliness, dilapidation, dirtiness, over-built space, and a lack of natural surroundings confirms the negative self-appraisal a person may have developed through other contacts with society. Self-esteem is the keystone to emotional well-being; a poor self-appraisal, among other factors, determines how one treats his surroundings and how destructive he will be toward himself and others. These factors set up a vicious circle that is difficult to break (Stainbrook, 1973).

How does the process of gardening enhance self-esteem? The gardener takes on a responsibility when he grows a plant. It is a living entity, its future dependent on the gardener's ability to provide conditions for growth. Each day as he tends his garden, the gardener observes the growth of his plants, and sees in that a measure of his success in planting, watering, and fertilizing. He anticipates and watches the slow but steady progress from seedling to young plant to full maturity and flowering. New leaves, stems, and flowers are his reward.

He identifies with his garden and builds a personal relationship with it. The garden becomes an extension of himself, a highly visible representation of his individuality. When it blooms he has brilliant evidence of his success. He also soon becomes aware that perhaps hundreds of people he does not know pass by each day and enjoy his garden. He has given them an anonymous gift. All of this enhances his self-image, helps to create self-esteem. The gardener, feeling better about himself, feels better about where he lives. The improved attitude about self and setting is evident in the gardener's comments and the unexpected non-horticultural benefits associated with these inner city gardens.

## QUALITIES OF PLANTS

What are the intrinsic qualities of plants that encourage people to respond to them? Plants are alive and are dependent on the gardener for care if they are to survive. In a world of constant judgment, plants are nonthreatening and nondiscriminating. They respond to the care that is given them, not to the race or the intellectual or physical capacities of the gardener. It doesn't matter if one is black or white, has been to kindergarten or college, is poor or wealthy, healthy or handicapped, plants will grow if one gives them proper care. They provide a benevolent setting in which a person can take the first steps toward confidence.

Plants communicate messages concerning life qualities to those who tend them. They display rhythms that are different from those of the man-built environment. Their growth is steady and progressive, not erratic and bizarre. The gardener sees a continuous, predictable flow of change from seedling to mature plant. He sees that change need not be disruptive but can be part of a dynamic stability. How different this is from our technological society, where the flow of life is constricted by schedule and regulation and must change rapidly to accommodate fads and other distractions, where people are under threat by new man-made terrors.

Plants take away some of the anxiety and tension of the immediate NOW by showing us that there are long, enduring patterns in life. It takes time for a cutting to grow roots, for a seed to germinate, for a leaf to open. Plants respond visibly to the sun in its daily course and signal the change of seasons. These rhythms in plants were biologically set in their genes by the same forces that set human biological clocks. An oak tree has looked like an oak tree for thousands of years. There is a certainty in knowing that a rose is a rose is indeed a rose—at all times and in all places.

Although gardening is usually pursued individually, it lends itself to group activity. Social interaction occurring in group gardening has benefits beyond horticulture that develop in response to the specific behavior setting. In the ghetto, for example, the garden provides a setting where people meet as neighbors to discuss growing of plants and other personal affairs. Through cooperative action, neighbors devise methods of protecting their gardens from vandalism. The garden becomes a catalyst for social intervention.

Matthew Dumont (1968), a community psychiatrist, has looked at the city to try to understand it in terms of the mental health needs of the city dweller. He states that city dwellers have a need for *stimulation* to break the monotony of daily life; for a *sense of community*, which arises, not because people are forced to live together, but rather from some spontaneous action such as creating a garden, and for a *sense of mastery of the environment*, reassuring them that they are not helpless cogs in the overwhelming machinery of living. Does not inner-city gardening speak to all these needs?

I do not mean to imply that the benefits of gardening are only available to residents of low-income neighborhoods. Rachel Kaplan (1983) surveyed over 4000 members of the American Horticultural Society, seeking the kinds of benefits and satisfactions they found in gardening. Over 80% of the survey participants placed "peacefulness and tranquility" in the top two categories of a five-point scale as the most important kinds of satisfaction gained from gardening. The answer of this more affluent group underscores the source of benefit as intrinsic to gardening. It is our humanity, not our economic or social status, that qualifies us to benefit from gardening.

## CORRECTIONAL INSTITUTIONS

The human values found in plants and gardening are effective in prison communities at correctional institutions. The programs, which range from farm work to organized instruction in horticulture and landscape design, not only break the boredom of confinement and provide vocational training, but in many instances have led to behavioral benefits. Maurice Siegler, former chairman of the U.S. Board of Parole, speaking of his experience as a warden at The Nebraska Penitentiary, said that although inmates might do violence to the buildings, they never destroy the plants they have grown.

Robert Neese (1959), while prisoner at the Iowa State Prison wrote,

These plants had a strangely soothing effect on our staff, when tempers did start to flare due to tension of constant confinement, a couple of hours of work in the garden made pacifists of potential battlers.

At the Massachusetts Correctional Institution in Bridgewater, the Assistant Deputy Superintendent commented on their horticulture program,

We feel that this is a remarkable program because it does so much for the men. It works out their tensions and as a result, there have been fewer assaults. The men who garden look forward to each season and can hardly wait till spring (Hunter, 1970).

In juvenile correctional institutions, gardening provides opportunities for creative group activities. Success with a vegetable garden can be particularly enriching for youngsters who have been unsuccessful in society.

## SCHOOL COMMUNITIES

Schools present another community setting in which gardening can be a source of education, creativity, and social well-being. The school grounds themselves are examples to the young people who pass through every day. Here children learn to enjoy and respect public spaces made for them, or to accept and reproduce litter-strewn, graffiti-covered, unkempt grounds (Primak, 1987).

In Chicago, long-term gardening projects at the Raymond, DuSable, and Austin Schools located in ghetto areas have shown their social benefit. Raymond School, an elementary school, started with an outdoor garden maintained by the students. Each class had its own flower bed, which the students designed, planted, and maintained. Each year since establishment of the garden, the number of broken windows in the school has been reduced. The garden has moved into the classrooms with a wide variety of nature and gardening projects in all grades. A pot of string beans on a widowsill provides an interesting math problem: How to divide six beans among thirty-five students? The school runs a plant sale that makes plants available in a plantless part of the city. At the Chicago Flower Show, all three of the schools participated with highly creative displays.

At a different kind of school community, Vassar College in Poughkeepsie, New York, trees play an important role on the campus, where a tradition of planting class trees started in 1868. Each graduating class provides funds towards the purchase of a tree and is guided by the Grounds Department in their choice. Plaques denoting the sponsoring class are present at each tree. The campus, heavily planted with trees, many of which are mature, of great size and character, is considered an arboretum. The trees provide a link between alumni and the college. During Reunion Weekend, many memorial celebrations are planned by the returning classes centering around the rededication and celebration of the class tree. The general appearance of the campus, its grounds and gardens, is high on the agenda of the Alumnae Council, which meets each Fall. President Fergusson commented,

> The Class Trees planted on the campus not only provide beauty for the beholder, they also create a link of alumnae/i who enjoy a return to Vassar in part to see their particular trees.

The trees have an importance for prospective students in choosing a school. The Carnegie Foundation for the Advancement of Teaching found that 50% of students and parents interviewed named campus visits as the most useful source of information when making their college choice. When asked what influenced them most during a campus visit, 62% said, "The appearance of the grounds and buildings."

## CONCLUSION

The interaction of people with plants can be of benefit in a wide range of community settings. We can only guess at the totality of the impact of plants in a human context—each of us has only a partial view. Can we accept the validity of this broader perspective? At universities, growth chambers and tissue culture labs can be seen as incubators for plants that ultimately will demonstrate the interrelationship of people with plants. The new plant forms produced by genetic engineering and plant breeding will resound in the human psyche. The products of scientific horticulture will finally be judged by their impact on the quality of everyday life of the people who plant and see them. An ancient bond, established during the co-evolution of people and plants, continues today, leading us to our examination of the role of plants in a human context (Lewis, 1979). We know so little, understand even less.

In their studies of the visual environment, Kaplan and Wendt (1972) have noted,

> If indeed man is likely to go out of his mind before he runs out of any of the traditional basic requirements of life, then a close look at the conditions and environments that are sanity-preserving and even satisfaction-enhancing seems essential. One facet of this issue is simply that of what man likes and why.

We know that the process of horticulture and its plants can indeed be sanity-preserving and satisfaction-enhancing.

In low-income housing, schools, prisons, and numerous other communities, plants and the growing of plants transcend the label *ornamental* to become an intimate part of creating a healthy human environment. Through a joint research initiative we can begin to unravel the ancient relationship of people with plants, to learn how it works, and quantify its effects in creating human well-being and promoting social development. Sharing these findings, we will enhance the quality of life for residents of this planet, our global community.

## LITERATURE CITED

Bonham, B. J., Jr. 1988. Green Towne Country: A development guide. Pennsylvania Horticultural Society, Philadelphia.

Bush-Brown, L. 1969. Garden blocks for urban America. Scribner's, New York.

Carey, L. J. (ed.). 1970. Community development as a process. University of Missouri Press, Columbia.

Dumont, M. 1968. The absurd healer. Viking, New York.

Gregory, R. L. 1966. Eye and brain: The psychology of seeing. McGraw-Hill, New York.

Hunter, N.L. 1970. Horticulture programs in prisons. Horticulture Department, California State Polytechnic College, San Luis Obispo.

Johnson, L. B. 1965. The conference call. In: Beauty for America: Proceedings of the White House conference on natural beauty. U.S. G.P.O, Washington, D.C.

Kaplan, R. 1983. The role of nature in the urban context. In: I. Altman and F. J. Wohlwill (eds.). Behavior and the natural environment. Plenum, New York.

Kaplan, S. and J. S. Wendt 1972. Preference and the visual enviroment: Complexity and some alternatives. In W. J. Mitchell (Ed.) Enviromental design: Research and practice. Proceedings of the Environmental Design Research Association Conference Three. Los Angeles.

Lewis, C. A. 1973. People-plant interaction: A new horticultural perspective. American Horticulturist 52(2):18–25.

Lewis, C. A. 1975. Nature city: Translating the natural environment into urban language. The Morton Arboretum Quarterly 11(2).

Lewis, C. A. 1979. Healing in the urban environment: A person/plant viewpoint. Journal of the American Planning Association 45:330–338.

Lewis, C. A. 1980. Gardening programs promote improved maintenance and community relations in public housing developments. Journal of Housing 37(11):614–617.

Neese, R. 1959. Prisoner's escape. Flower Grower 46(8):39–40.

Newman, O. 1972. Defensible space. Macmillan, New York.

Primak, M. (ed.). 1987. The greening of Boston, an action agenda. Report from the Boston Foundation Carol R. Goldberg Seminar. The Boston Foundation, Boston.

Stainbrook, E. 1973. Man's psychic need for nature. National Parks and Conservation Magazine. 47(9):22–23.

CHAPTER 9

# Philadelphia Green's Greene Countrie Towne Model as an Agent for Community Development

## Findings of Case Studies

---

J. Blaine Bonham, Jr.

Director, Philadelphia Green

## INTRODUCTION

Philadelphia Green, the community outreach program of the Pennsylvania Horticultural Society, works with residents in the city's low and moderate income row house neighborhoods to plan and carry out greening projects. Projects include turning vacant lots into vegetable and flower gardens, and lining streets with trees, shrubs, and flowers—all to create a physical transformation of the neighborhood through the efforts of residents working together with Philadelphia Green. Over the last fifteen years, the program has sponsored about 1200 projects, and its success is evident from the long waiting list of groups applying to be part of the program.

The current strategy at Philadelphia Green is to develop Greene Countrie Townes. (This appellation was used by William Penn in 1682 to describe his City of Brotherly Love, where he sought to leave one acre cultivated for every 10 developed.) We have identified neighborhoods that have a strong track record of gardening projects. Philadelphia Green tries to locate a community-based umbrella organization in each neighborhood to assist in disseminating information, organizing blocks around gardening projects, and helping with the overall logistics. Over a three- to five-year period, we concentrate many resources in that neighborhood to green-up city blocks, corridors, gateways to the neighborhood, and visual and social focal points. To date, we have dedicated four Greene Countrie Townes and three more are under development. They have ranged in size from 8 blocks to 200 blocks. Within larger neighborhoods, we identify greening clusters where interest and enthusiasm run particularly high.

## PHYSICAL IMPACT

The main goal in the Greene Countrie Towne is to effect an improvement in the physical conditions in the neighborhoods. To this end, the program has been exceptionally successful. In some Townes, most of the vacant lots have been made into spots of beauty, sources of food, and causes for pride. The hard edges of the built environment have been softened with tree-lined streets and rows of containers with flowers and shrubs. Plants make the street, the communal space in the neighborhood, a more livable place.

Many key sites throughout these neighborhoods serve as community show pieces: the Wedding Garden in Point Breeze, with its Victorian gazebo, surrounded by a wrought iron fence; Memory Lane, the sixty-house celebration site in Francisville, with its six connecting gardens, its tree park, and "sculpture" garden; Mt. Kilimanjaro in Susquehanna, its dramatic African scene of that mountain setting off the garden with ornamental grasses and perennials; the four-acre Green Acres vegetable and hillside garden, complete with the "ribbon of gold" of wildflowers. The Greene Countrie Townes visually serve as triumphant examples of a community-oriented program that involves people integrally in the success of the process. They serve ultimately as testaments to the importance of plants in the dense urban environment.

## SOCIOLOGICAL BENEFITS

The Greene Countrie Townes have also produced sociological benefits, a bonus to the physical transformation. With new Greene Countrie Townes, we now aim for several sociological goals as well:

- Increased participation in greening: By virtue of the program's success, thousands of people across the City have assumed responsibility for the improvement of their neighborhood, both at the time of project initiation and in their commitment to long-term maintenance. In the Strawberry Mansion neighborhood, tradesmen are devoting their time and skills to building patios for gardens; in Francisville, they are building barbecue pits and benches. The Susquehanna neighborhood is witnessing the emergence of four major children's gardens, all with adult leaders, thus involving this vital group into the greening process. A well-maintained garden is a sign of the value placed on it by everybody on the block.

- Increased leadership skills: Across all the neighborhoods, people who have not previously been involved in a leadership capacity have emerged. These garden leaders assume responsibility for organizing the neighbors to carry out the project and for maintaining the project. Also, Philadelphia Green has a 50-person advisory board of these community leaders, many of whom serve as volunteer chairs for our gardening events.

- Increased organizational capacity: Both at the block level and the umbrella organization level, experience and success can enable a group to do more. The non-profit Point Breeze Federation, a social service agency in that neighborhood, used greening as a way of bringing member blocks together and creating a community-wide awareness about Point Breeze. The Federation has received grants to hire an adult lead gardener for the neighborhood and to hire youths in summer to help senior citizens maintain sites. The West Shore community has held annual plant sales to raise money and encourage greening. Green Acres in Susquehanna is building a more broadly based committee structure for its 80 gardeners through a series of structured workshop activities.

- Other community development activity as spin-offs: Many neighborhood leaders claim that the success of the Greene Countrie Townes has led residents to want to do other things in the neighborhood, from clean-ups to town watches, and has attracted private and federal funding to them for greening and other projects. West Shore attracted a national housing program, which has helped to renovate 25 of 40 vacant houses. Point Breeze attracted federal dollars for a summer youth program, money to renovate its community office, and for scattered site low-income housing.

## A STUDY TO ASSESS OUR EFFECTIVENESS

Admittedly, these sociological goals were not part of the original intent of the Greene Countrie Townes when this focus first began ten years ago. Yet, they have undeniably occurred, in varying degrees. As part of our long-range planning process, we commissioned a study in July 1989 conducted by the Organization and Management Group to examine the factors that affected our ability to achieve these sociological goals and the implications for future programmatic approaches and development.

The study has focused on three Greene Countrie Towne communities: Point Breeze, Francisville, and Susquehanna. The Point Breeze study concentrated on management and leadership dynamics in relation to the quality of on-going maintenance. It also examined the effects of the greening model on the major organizations and the generation of other community development activities conducted. The major methods in this part of the study included participation in garden-related activities and interviews with staff, community leaders, and gardeners.

The last two parts of the study relied on interviews with staff, community leaders, and gardeners. The consultants also reviewed relevant statistical and planning documents of the program and city government agencies. The Francisville study focused on the evolution of program planning strategies and their relationship to the overall community power structure. The Susquehanna study looked at the effect of program planning on the success of individual sites.

### Key Findings

The investigation has produced preliminary findings that are grouped under three major issues raised by each of the case studies as they relate to Philadelphia Green's social goals of community development through Greene Countrie Townes.

**Issue 1.** The effects of management, leadership dynamics, and the indigenous neighborhood power structure on the ability to achieve sociological goals.

*Findings.* The Green Countrie Towne Program has differing effects on the various levels of organizational structure. Sometimes the effects among these levels within a particular community are correlated; other times, the effects are mutually exclusive. The Program can help develop the capacity of community-wide institutions and help to leverage other community development projects without affecting the leadership dynamics of individual blocks or the gardens themselves.

The management and leadership dynamics of the gardens play an important role in the promotion of social community development goals. Organizational "democracy" is the exception; benevolent monarchy the rule. The average gardener focuses on his/her plot and is not interested in overall management or general maintenance. This situation limits the development of additional leadership skills.

The power and leadership structure of the umbrella community organization plays a major role in the promotion of Philadelphia Green's goals. The importance of establishing a working relationship based on cooperative understanding of roles, process, and working styles between Philadelphia Green and the neighborhood leadership is crucial to an effective program. Powerful community groups can be unwilling to share their power base, thereby effectively limiting the degree to which program staff can promote interest and participation.

A leadership crisis is pending. Most of the garden leadership in all three communities is over age 65. In many cases, no succession leadership is apparent. There is a great need to develop interest among younger people.

**Issue 2.** The implications of the demands of quality maintenance on the management/leadership structure and impact on self-sufficiency.

*Findings.* The physical condition and maintenance of garden projects does not correlate with the quality of the organizational and management structure. General maintenance of communal areas was accomplished in some cases by one or a few people. Most gardeners focused their attention on their individual plots.

Many gardening groups have become overly dependent on the program for maintenance, resources, and services. A more clearly stated joint contract concerning the expectations of assistance is needed. Older gardeners genuinely require more maintenance assistance.

The more direct assistance that program staff provides gardeners in the form of maintenance, the more gardeners throughout the whole community expect Philadelphia Green to provide maintenance, resources and services, thereby decreasing their self-sufficiency.

The drug epidemic has greatly influenced some gardeners' maintenance levels as the possibility of assault and/or vandalism has substantially increased.

**Issue 3.** The effects of program planning practices on the achievement of social goals.

*Findings.* It is crucial for the staff to understand how to work effectively with various neighborhood power structures. In some cases, we lacked this understanding, sometimes working around the indigenous structure. This complicated the process and made the achievement of overall Greene Countrie Towne development goals more difficult.

As a distributor of resources, Philadelphia Green becomes a political player in both community-wide and garden politics. Within a neighborhood, various gardening groups see themselves as competing for program resources. Also, some powerful umbrella groups viewed our program as a threat to their control. Largely, Philadelphia Green field staff was unaware of this role in the community power structure.

Philadelphia Green's vision of good leadership and management in some cases differed substantially from the community's goals. Our goals are sociological, as well as physical. As a result, gardeners often resist the efforts of staff to install a more organized "democratic" structure, especially in large gardens.

The staff's desire for beautiful gardens must be balanced with the need for self-sufficiency of the gardeners. When the staff personally performed various maintenance tasks, dependency increased and self-sufficiency was not fostered.

## NEXT STEPS FOR PHILADELPHIA GREEN

Philadelphia Green has a deep commitment to helping Philadelphia's residents revitalize their neighborhoods. As a consequence of this study, Philadelphia Green continues to strengthen its commitment by beginning to revise certain policies and practices with the aim of becoming more effective in developing community organizational capacities. By becoming better at fostering this self-sufficiency, Philadelphia Green believes the horticultural improvements that neighborhood groups make to their community will be longer lasting.

In planning for the next several years of program development, we have begun to incorporate the following in our approach:

- With the current Greene Countrie Townes, develop a year-round program of horticultural/technical assistance and organizational workshops to build capacity and self-sufficiency within the neighborhoods around their greening efforts.

- With future Greene Countrie Townes, develop agreements with the neighborhood leaders, provide initial training to them before site development begins, and jointly establish guidelines for cooperation.

- Establish new strategies for allowing gardens to develop to a level according to the capabilities and commitments of the residents themselves.

- Train the field staff to become more knowledgeable at assessing and working with the existing power structures in the community.

- Develop a youth program based in the Greene Countrie Townes to involve kids in neighborhood greening activities and foster an appreciation of their neighborhood environment.

CHAPTER 10

# A Research Agenda
# for the Impact of Urban Greening

---

Mark Francis

Professor and Director, Center for Design Research, University of California, Davis

Chris Cordts

Coordinator, Urban Gardening Program, Colorado State University,
Denver Cooperative Extension

Board Members

American Community Gardening Association

The American Community Gardening Association (ACGA), in its efforts to improve the knowledge and practice of community greening, is developing a research agenda. The agenda consists of critical questions and issues requiring continued research, innovation, and testing. An expert panel [1] assisted ACGA in refining and expanding the agenda and the Board reviewed and identified items of priority concern.

The agenda will be used to inform the ACGA board and its members as well as researchers and designers interested in urban greening projects. This list is seen as additive and changing, depending on the needs of practitioners and public officials involved in greening activities such as community gardening, community open space development, urban horticulture, and urban forestry in American towns and cities. Comments and suggestions are welcomed as the agenda will be revised and updated according to ACGA concerns and changing trends in community greening.

The research agenda is organized around several major categories, including individual, group, economic, etc. effects of community greening. The top ten ranked issues by

ACGA Board members are listed below. Items marked in bold are ones identified as priority areas by five or more ACGA Board members in their ranking.

## RANKING OF TOP TEN AREAS FOR RESEARCH (Priority 1/2 ranking by ACGA Board)

1. Effect on property values (6/2)
2. Relationship to City Wide Open Space Policies & Plans (6)
3. Participatory Design & Planning (5/2)
4. Community Gardening as Individual Empowerment Tool (5/1)
5. a. Constituencies for Community Greening (5)
   b. Contribution to Building Social Cohesiveness (5)
7. a. Meeting Place for Different Cultural Groups (4/2)
   b. Definition of Community in Community Greening Projects (4/2)
   c. National Policy and Programs for Community Greening (4/2)
   d. Relationship of Land Ownership to Project Permanency (4/2)

[Bold indicates selected by 5 or more ACGA board members as a priority issue.]

### Individual Benefits

- Use/nonuse/misuse of community greening projects
- Importance of projects in providing "nearby nature" for urban residents
- **Developmental benefits for children and youth, elderly, etc.**
- Accessibility
- Effect on environmental learning and competence
- Environmental autobiography/childhood memories of gardens, community greening projects
- Importance of "familiar place" in an unfamiliar environment
- **Community greening as individual empowerment tool**
- A place to exert individual control
- Recreational benefits
- **Effect on reduction of stress**
- Health/healing benefits
- Effect on "restorative experience"
- Preference for "built" versus "natural" elements
- Source of fresh, nutritious food
- Workplace satisfaction for nearby office workers/employees

### Group/Neighborhood Benefits

- Contribution to public life and culture
- Setting for social interaction
- **Impact on group identity**
- **Contribution to building social cohesiveness**
- **Meeting place for different cultural groups**
- **Cultural differences in use**
- A setting to expose and resolve social conflict
- Cultural expression
- **Definition of community in community greening projects**
- Role in urban neighborhoods
- Role in suburban neighborhoods

### Political Aspects

- **Greening as a political force**
- Empowerment of the disenfranchised via growing/working with plants
- Role in political exchange and education
- **Constituencies for community greening**
- **Relationship of greening to the environmental politics of 1990s**

## Security/Safety

- Impact on crime/crime statistics
- **Vandalism or lack of due to gardens**
- Garden protection
- Sense of ownership

## Aesthetic Benefits

- Effect on landscape perception
- Visual assessment of urban greening projects
- Visual diversity and complexity

## Sense of Place/Benefits of a Meaningful Landscape

- Meanings
- Symbolic importance
- **Sense of place**
- Sense of stewardship

## Philosophical/Theoretical Aspects

- The role of nature and community greening in public life
- Ecological diversity
- Typologies of projects
- Community greening projects as social and political symbols

## Temporal Qualities

- Sounds
- Smells
- Touch
- Seasonal differences
- Use of senses and gardens for multiply handicapped people

## City-wide/Regional Benefits

- Air quality
- Role in city-wide and regional identity/attachment
- **Relationship to city-wide/regional greening and greenway programs**

## Ecological Benefits/Impacts

- **Functions as part of larger ecological systems**
- Impact on global warming
- Relationship to urban forestry programs and techniques
- **Habitat for birds and animals**
- Effect on air quality
- Effect on recharging of water table
- Water and energy utilization
- Recycling benefits
- Soil quality and toxicity
- Role of plants in exploration of space

## Economic Benefits/Impacts

- **Development costs**
- **Maintenance costs**
- Relationship to tourism
- **Effect on property values**
- **Relationship to adjacent housing projects**
- Effect on retail sales of adjacent commercial properties
- Effect on occupancy rates/rental rates of adjacent office properties
- **Greening as job training**
- **Economic impact of urban greening on green industry**
- **Value of food produced**
- Contribution to city revenue/tax base

## Horticultural Aspects

- Plant care
- Arboriculture
- Planting techniques

## Historical Aspects

- **History of community gardening**
- Historical shifts
- Political history of community greening

## Design and Planning Issues

- Role of the design professional
- **Participatory design and planning processes/impact on empowerment**
- **Teaching of community greening in design and planning schools**
- Importance of project size to success
- Relationship to overall community design
- Relationship of project form and style to project success
- Design patterns
- **Integration of projects into new and existing housing projects**

## Permanency/Ownership Issue

- **Ownership as a strategy for permanency**
- **Best type of ownership from gardeners' perspective**
- Community gardens as part of Park District system

## Public Policy Issues

- **Relationship to city-wide open space policies and plans**
- **National policy and programs for community greening**

## Permanency

- **Relationship of land ownership to project permanency**
- Impact of land trusts

## Management and Maintenance

- **Successful maintenance programs**
- **Successful project management techniques**

## Sustainable Development/Agriculture

- Relationship of community greening to sustainable agriculture

## Methodological Issues

- **Quantitative measures/methods**
- **Qualitative measures/methods**
- Useful/low-cost evaluation strategies for local groups
- Action-research approaches
- Research funding

## NOTE

[1]The panel included Gary Appel (Life Lab Science Program), Lisa Cashdan (Trust for Public Land), Stevie Daniels (Rodale Institute), Charles Lewis (Morton Arboretum), Douglas Patterson (University of British Columbia), Diane Relf (Virginia Polytechnic Institute), Herb Schroeder (US Forest Service), W. Gary Smith (University of Delaware), Robert Sommer (Center for Consumer Research, University of California, Davis), Roger Ulrich (Texas A & M), and Sam Bass Warner, Jr. (Boston University).

CHAPTER 11

# The Role of Nature for the Promotion of Well-Being of the Elderly

Charlene A. Browne

Assistant Professor of Landscape Architecture,
Virginia Polytechnic Institute and State University

## INTRODUCTION

Senior citizens represent the fastest growing segment of the American population. As a result, retirement communities within the United States are proliferating at a rapid rate. Many of these communities have landscaped grounds and outdoor amenities; however, most have not been assessed regarding the impact these landscaped settings have on the residents. This paper reports findings from a research project, funded by the National Endowment of the Arts, which has been directed at understanding the extent to which outdoor settings within retirement communities promote psychological, social, and physical well-being. Specifically, this paper will address five areas in which nature[1] may have an impact on the promotion of well-being: aesthetics, environmental stimulation, social interaction, motivation for physical exercise, and self-expression.

The author utilized a multimethod research approach including twelve site visitations (indicative post-occupancy evaluations), two questionnaires (one administered to the management and the other to residents), and interviews with selected residents of four retirement communities to determine their outdoor visual and spatial preferences.

## GENERAL FINDINGS

### The Importance of Pleasantly Landscaped Grounds Cannot Be Overestimated

According to 99% of those who responded to the questionnaire,[2] "living within pleasant landscaped grounds within their retirement community" was considered to be either essential (50.5%) or important (48.5%). When asked which outside features they felt were important to have within a retirement community, 95% of the residents thought windows facing green, landscaped grounds were either essential (35%) or important (60%). Although views from windows to activity spaces were also valued, only 6% said views facing outdoor activities were essential, and 34% said important. Interviews with 67 residents from the four communities support this preference for "greenscape."

When asked in an open-ended format, "Why did you select this particular retirement community?" the following responses (201 total) were given: family and friends, 24%; grounds/environment-configuration, including the site plan, scale, and general character of the grounds, 24%; grounds/environment-nature, including the naturalistic character of the grounds such as water, vegetation, and views to nature, 14%; services, 15%; setting/surroundings, 14%; religion, 8.5%; and economics, 2.5%. These responses reveal that the character or quality of the grounds, as well as the inclusion of nature within this type of environment, is important to residents. If the second and third categories are combined, the concern for the configuration and natural elements of the grounds represents 38%—a large percentage of responses.

Responses from the interviews support this preference for nature. When asked to identify their favorite places, residents preferred the following landscape elements: a body of water with a naturalistic character, panoramic views and open green spaces, and trees that enclose a large open space or have an ornamental quality. It appears that residents prefer a "naturalistic" landscape, or "softscape," over a "hardscape," that is, a predominantly paved area.

### Outdoor Spaces that People Enjoy Using Reflect an Interest in Nature and Nature Observation

Outdoor activities tend to be informal, passive, and reflective in character and performed by a few people at a time. When asked which outdoor activities residents enjoyed in general, not necessarily within their retirement community, the following responses were given: walking, 85%; enjoying nature, 62%; talking with friends, 62%; looking at plants, 46%; and gardening/garden plots, 21%. Residents responded similarly when asked to identify their favorite outdoor activities or amenities within their community: walking paths, 62%; own patio, 58%; lake, 39%; informal gardens, 32%; and garden plots, 21%.

Although the responses of management and residents were similar, the management underestimated both the residents' preferences related to the patio as an intimate place that provides some privacy in the out-of-doors, and the importance to the residents of observation of plants, water, and animals.

### Five Areas in which Nature Can Have an Impact on Well-being

**Psychological Well-being Through Aesthetics.** Plants and overall landscape architectural design help to create an aesthetically pleasing, home-like, non-institutional setting, as opposed to a medical or institutional setting. Outdoor spaces should have a purpose, be of human scale, and provide a variety of experiences. Garden settings; nature observation areas; and variety, color, and detail in vegetation are key characteristics. Some residents are unable to spend much time outside due to poor health or mobility. For those people, views of nature become very important. Residents (especially those who are less mobile) prefer a superior viewing position and panoramic views with long vistas framed by plants, in an informal setting with water, grass, and trees. Also, neatly trimmed plants that provide spatial order and legibility seem to be preferred.

**Environmental Stimulation.**   Retirees spend much of their time at home and their schedules usually are not prescribed as in the work world. They often become accurate and acute observers of their environment. Therefore, it is important to provide the campus with a variety of plants that have seasonal variation. This aspect of change is not only aesthetically pleasing, but also triggers pleasant memories of their previous home environments, helps to maintain their mental activity and awareness of time, and decreases boredom—a common situation among the elderly. Plants attract and provide shelter for wildlife, as well as provide shaded resting areas for nature observation, a key activity of the residents. At the retirement communities, bird feeders are frequently attached to trees and posts. Water, a natural element preferred by residents, is also an important natural element for attracting wildlife. Lakes, rivers, and lagoons become attractors for a variety of migratory water fowl and provide homes for introduced species, such as swans. The swans provide many hours of entertainment for the residents as they observe these birds progress through various aspects of the natural life cycle: the building of their nests, nesting on the eggs, and the growth of their young. Not only do these kinds of observations keep residents in touch with segments of time and seasons, but also with the natural process of the life-death cycle, which is inevitable for us all.

**Self-expression and Personalization.**   In three of the four communities, residents in the cottages are encouraged to select their own plant palette, as well as to plant and maintain their own gardens within approximately four feet of their structures. For the most part, this is a very successful practice on a variety of levels: it provides for self-expression, individuality, and nurturing; it keeps people active and interested in their own home; it encourages pleasant conversation about the plants with those passing by; it is very aesthetically stimulating due to the variety of tastes reflected in the choice of plants; it is enjoyed by residents using the walking paths; and it provides for a home-like, non-institutional atmosphere for living.

**Motivators for Physical Exercise.**   As these findings suggest, walking is the most popular outdoor activity among retirement community residents. Plants, if properly utilized, can encourage residents to participate in the out-of-doors, as well as to reap the benefits of physical exercise. It is important for the landscape architect to consider the various design elements of the path that enhance leisure activity. When asked to identify their favorite walking route, residents from both communities in Florida preferred routes that had water as a primary attraction. Their secondary responses related to the variety and beauty of the plants on the site. In the Florida communities, people preferred plants with large flowers of red, hot pink, yellow, and violet colors, such as the Royal Poinciana, the Princess crepe myrtle, oleander, and orchids. Fruit trees such as mangos, bananas, oranges, grapefruits, and star fruit, and plants of unique or exotic qualities such as the Saeko Palm, Banyan Tree, or the Bird of Paradise were also identified. The residents interviewed at the two communities in Pennsylvania preferred similar characteristics of plants, although different species were used. Woods consisting of Beech, Oak, and Tulip Poplar were popular. Several trails within the forest are maintained and used on a regular basis in the summertime. Brightly colored annuals, perennials, and roses were popular among all four communities. Neither the scent nor tactile quality of the vegetation was a major factor in the popularity of the plants—visual appeal was the primary concern. The use of plants with diverse, unique, and seasonal qualities along walkways may encourage people to explore the out-of-doors to a greater extent. Trees that provide even shade should be used, as dappled shade may create patterns on the sidewalk that some of the elderly find difficult to interpret. Interesting plant assemblages with benches should be placed at intersections of the pathways to encourage greater use.

**Social Interaction and Networking.**    Community garden plots, strategically located along walking paths, provide a focus for conversation and interaction. Residents who may not be able to participate in gardening due to health are able to talk with and observe those who can. The former can observe the growth of vegetables and flowers, and give helpful suggestions from their previous gardening experiences—this association often triggering fond memories for them. Selection of plants for the community campus can become a group affair, whereby residents, after careful consideration, select plants for specific locations. Also, it is suggested that trees have name tags so, to a degree, the campus becomes an arboretum. Management's encouragement of this kind of group interaction and decision-making process enhances social interaction and networking among residents and staff.

## SUMMARY

As is evident from this study, landscaped grounds and the inclusion of natural elements are important to residents living within the retirement community setting and may play a role in promoting psychological, social, and physical well-being among residents.[3] Research with regard to the visual and spatial preferences of the elderly in relation to the design of outdoor environments is minimal at best. In the literature that is available, Carstens (1985) suggested that each community is unique and has different preferences in regard to the programming for outdoor spaces and activities. The author found, however, that specific landscape elements are preferred among a variety of retirement communities throughout the United States. These preferences include outdoor spaces that support informal and passive activities such as walking and talking with friends; gardening; nature observation of water, vegetation, and fauna; and panoramic views of a naturalistic softscape. As the residents' world becomes smaller due to less environmental options as a result of loss of independence and mobility, a changing and expansive panoramic view becomes important. Therefore, outdoor spaces that support these preferences should be a priority over other elements that are not used as frequently. Also, the underestimation of the importance of nature and related activities in retirement settings is evident in the fact that plants and other aspects of nature are underutilized or not utilized appropriately for maximum benefit.

Some suggestions have been given to improve this situation; however, additional research related to nature and the promotion of well-being among the elderly is needed, since the concern for creating a life-sustaining and life-enriching environment for this group is an important issue facing society today.

## NOTES

[1]"Nature" in this context is defined in a fairly broad context and includes plants, vegetation, and fauna. It also includes landscape settings with plants whether they are within a natural or man-made setting. The term "naturalistic" as used here denotes "softscape" as opposed to "hardscape" areas with a predominant amount of paving.

[2]A questionnaire was sent to a random sample of 30% of the residents (minus the nursing care patients) within nine retirement communities. 680 questionnaires were mailed with a 55% return rate. Also, interviews on the visual and spatial preferences were conducted at four communities.

[3]Rachel and Steven Kaplan have provided a strong conceptual framework in which to understand the impact that nature may have on psychological well-being. Suggested reading includes Kaplan & Kaplan (1989).

## LITERATURE CITED

Carstens, D. Y. 1985. Site planning and design for the elderly: Issues, guidelines, and alternatives. Van Nostrand Reinhold, New York.

Kaplan, R. and S. Kaplan. 1989. The experience of nature: A psychological perspective. Cambridge University Press, New York.

CHAPTER 12

# Children's Gardens: The Meaning of Place

Catherine Eberbach

Manager, Children's and Family Programs, New York Botanical Garden

I recently overheard one our Children's Garden instructors tell a group of five-year-olds that they were going to walk through the garden and look for signs of spring. One boy, obviously distressed, cried out, "But I can't read!"

You and I know that the teacher intended to find daffodils, dandelions, and other harbingers of spring. But from this child's limited experience, signs were something else entirely. Perhaps he thought of street signs, which require reading skills to be understood. Whatever he thought, his reaction erupted from the context of *his* experience and the meaning *he* gave to the word.

Just as our perceptions shape the meanings we give to words, our perceptions also give meaning to "place." This example illustrates that adults and children see things differently. As with words, children define and experience place differently than adults; my research and work with young people suggests they also have their own perceptions of gardens.

Five years ago I undertook a graduate project to design a garden for children at Longwood Gardens (Kennett Square, Pennsylvania). Determining children's preferences was primary to creating an interactive garden: How do children think of gardens?; What are their favorite garden features and elements?; What do they want to do in a garden? In short, what does a garden mean to a child? To begin addressing these queries, I collected drawings of gardens from 178 elementary school students, each of whom described his/her drawing. [Refer to Eberbach (1987, 1988) for a detailed study.] I selected artwork because research indicates that children draw what they remember, and they remember the things that they value (Lark-Horowitz, et al. 1967).

This study and subsequent work with youth gardening led me to several observations, three of which are relevant to this meeting. To begin, these drawings indicate that children understand the essence of a garden to be a plant environment. Ninety-eight percent of the drawings included plants in one form or another—from seeds, to colorful flowers, to people-eating trees. The 2% who graphically omitted plants, included them in verbal descriptions. One student described his drawing of brown scribbles as the soil where "plants are growing, but you can't see them."

Children also expressed preference for plant qualities. Overall, 47% chose ornamental plants (i.e., non-edible plants selected for aesthetic purposes). Compare this with 19% whose gardens were exclusively of functional plants such as fruits, vegetables, and herbs. This finding contradicts the popular notion that children are interested in gardening activity to the exclusion of a garden's appearance. In fact, children *do* care how gardens appear; they *do* make aesthetic judgments, although this observation does not assess the merits of these judgments.

In this study, first graders preferred functional plants and fourth graders preferred gardens that combined ornamental and functional plants. These deviations from the norm lead to my second observation: How a child experiences and perceives a garden is a function of development.

I would argue that cognitive abilities shape a child's perception of gardens. The thinking of young children is centrated, that is, focused on a few environmental elements that they cannot integrate into a unified whole. This is obvious in the drawings by first graders, which are generally limited to plants and soil. Moreover, these drawings are characterized by rows of isolated objects, a sort of list. It seems likely that children at this developmental level value the parts because they are unable to perceive the whole.

As cognitive abilities develop, thinking becomes decentrated, that is, many and varied environmental components are perceived and can be linked into a whole picture. This is demonstrated in drawings of older students that portray gardens as no longer solely made of plants, but as places with animals, water, paths, tools, people, and even tennis courts. These items are not merely parts of a list, but are parts of a design.

Cognitive functioning is one explanation of why the drawings of first graders are dominated by single, basic elements: They simply lack the ability to do otherwise. It would also explain why fourth graders integrated ornamental and functional plants into their drawings: greater cognitive maturity enables them to link many different environmental components.

Exclusion and inclusion of paths may also be an indicator of cognitive ability. Paths are strikingly absent from the drawings of first and second graders, but are prominent in those of fifth graders. If cognitive development is considered, this is to be anticipated:

> Younger children are just developing cognitive skills that are necessary to negotiate pathway systems and reverse routes. For this reason, it is unlikely that pathways would be included in their art-work. As children develop cognitively, they use pathways more expertly, and this is reflected by a higher incidence of pathways in their drawings (Eberbach, 1987).

Cognitive development is linked to the third observation: Children use activity to define and give meaning to gardens. This is especially significant in view of how children actually learn. According to Piaget, cognitive development is facilitated through hands-on, active participation. Children learn to think logically, get along socially, and even resolve moral dilemmas through doing, doing, doing. Abstract problem solving remains a distant skill. Just as activity is a child's modus operandi, it is also used to define space.

These drawings are saturated with activity—moving elements like animals, water, and people enliven otherwise static scenes. Other elements imply activity—paths to walk, bridges to cross, swings to swing, ladders to climb, and tools to garden.

Listening to descriptions of their drawings, and in subsequent conversations with children, I am often struck by how consistently gardens are discussed in terms of activity. For example, here is how one boy described a drawing of his ideal garden: "These are bushes where I can hide . . . and this is a treehouse we built with stuff we found. . . ." Others described gardens as places where they would read, do homework, play, escape younger siblings, or invite special guests. Furthermore, the drawings of those who gardened in our Children's Garden Program included themselves in their drawings of gardens—a powerful statement about the impact of child participation.

Contrast this with how adults might describe a place. How would you describe this room we are in now? As a place where people sit and listen, while others talk and show slides? Or would you think in terms of dimension and design? Take this a step further and consider how adults describe gardens. In her book *Gardens of the Italian Villas* (1987), Marella Agnelli writes the following about Villa Reale:

> From the facade of the villa itself stretches a huge lawn enclosed at the far end by a semicircle of tall boxwood. A garden room containing a magnificent fountain and grotto leads to an area planted with lemon trees . . .

This is clearly not a child's description. It conjures up an image by detailing layout without creating a sense of one's own involvement. Perhaps then, a child's use of activity to describe a garden is really a way to understand his/her place in the garden.

If gardens are to be relevant to young lives, children's perceptions and preferences must be considered, particularly as children can specify preferred plant qualities and garden elements. What are the characteristics of a well-designed garden for children? At the very least, such a place addresses the three observations presented today:

- children understand what gardens are and have aesthetic preferences;

- perceptions of gardens are shaped by a child's cognitive development; and

- activity is used to understand a garden and the child's place in the garden.

How can these observations be applied to the design and use of garden space for children? The first step is to listen to what kids have to say about color, plant qualities, and preferred activities. It is then the designer's challenge to interpret these opinions into a suitable design. Bringing youth into the design process, though challenging, has many benefits. They are more likely to assume ownership, which will be reflected in their use and enjoyment in the garden.

Secondly, child development is a pivotal issue when designing children's gardens. At The New York Botanical Garden (NYBG), one way our Children's Garden recognizes developmental skills is through the design of its individual plots. Five and six year-old gardeners tend 4' × 4' wood-framed plots that are raised a few inches above ground to help distinguish between garden and path. Because older children can make this distinction and are physically stronger, their plots are not framed and are 4' × 15'.

Above all else, highlight activity. This is the cornerstone of the Children's Garden at Longwood; youngsters play with the fountains, touch plants, interact with topiaries, crawl through plant-laden tunnels, climb bridges, and hide behind maze walls.

Not all activity need be so costly, or so play-directed. The layout of NYBG's Children's Garden accommodates many other kinds of activity. In addition to planting, weeding, and harvesting, kids can find their way along the paths of a maze made of vegetables and flowers. We discovered that it really does not matter that children can see where all paths lead; they still pretend to be lost and to hide behind plants only 10 inches high. During 1990, our young gardeners designed and built a pond and now actively study its progress. A meadow and piles of compost and manure are favorite spots where kids congregate and play spontaneously.

These examples only begin to illustrate how children's gardens can reflect the interests and pleasures of children. So much more remains to be discovered . . . How do children use garden spaces? How does playing in gardens influence child development? How do experiences in gardens shape a child's view of natural environments? With greater attention to children's perceptions and preferences, a child's experiences in the garden can only be richer and vastly more rewarding.

## LITERATURE CITED

Agnelli, M. 1987. Gardens of the Italian villas. Rizzoli International Publications, NY.

Eberbach, C. 1987. Gardens from a child's view: an interpretation of children's art-work. Journal of Therapeutic Horticulture 11:9–16.

Eberbach, C. 1988. Garden design for children. MS Thesis, University of Delaware, Newark.

Lark-Horowitz, B., H. P. Lewis, and M. Luca. 1967. Understanding children's art for better teaching. Charles E. Merrill Books, Inc., Columbus, OH.

CHAPTER 13

# Socio-Economic Impact of Community Gardening in an Urban Setting

Ishwarbhai C. Patel

County Agricultural Agent, Urban Gardening Program,
Rutgers Cooperative Extension

## INTRODUCTION

Community gardens are neighborhood open spaces managed by and for the members of the community. Most typically, the community garden is divided into individual plots and planted with vegetables by landless gardeners. Some families even share the same plot. Each community garden is a smorgasbord of gardening methods and styles. One will see raised-beds, flat-beds, mulched gardens, herbs, organic gardening, vertical gardening, succession gardening, square-foot gardening, and plant varieties of many ethnic heritages.

Although many benefits can be derived from community gardening, only the socioeconomic effects are discussed here. The individuals and families living in Newark, New Jersey and surrounding communities provide the basis for our research (Green, 1981, Program Staff, 1989a,b).

## SOCIOECONOMIC TRAITS OF GARDENERS

The data in Table 1 reveal that almost two thirds (65%) of the gardeners were women and 35% were men; most (75%) of the gardeners were black and 5% were white. Hispanics accounted for 19%. One teenager (1%) was represented in the sample, but most gardeners were middle-aged (56%) or senior citizens (43%). Over two fifths (44%) were high school graduates, one tenth (10%) were college graduates, and about one fourth (26%) had received education to the elementary school level. Over one third (37%) were gainfully employed and the rest were retired (34%), disabled (13%), or welfare citizens (6%).

**Table 1.** Socioeconomic traits of gardeners[a].

| Trait | | Percent | Trait | | Percent |
|---|---|---|---|---|---|
| A. | Sex | | D. | Formal education | |
| | Men | 35 | | Elementary | 26 |
| | Women | 65 | | Junior high school | 16 |
| | | | | High school | 44 |
| B. | Race | | | College | 10 |
| | Black | 75 | | Other | 4 |
| | Hispanic | 19 | | | |
| | White | 5 | E. | Source of livelihood | |
| | Other | 1 | | Self-employed | 12 |
| | | | | Employed by others | 25 |
| C. | Age | | | Retired | 34 |
| | Adults | 56 | | Disabled | 13 |
| | Seniors | 43 | | Welfare | 6 |
| | Youth | 1 | | Other | 10 |

[a]Survey results based on personal interviews wtih 45 gardeners in 1980 and 133 gardeners in 1987.

## SOCIOECONOMIC BENEFITS OF GARDENING

The gardeners cited many benefits that they believed resulted from gardening. Table 2 identifies these perceived benefits.

The majority of benefits reflects the value of horticulture to human well-being. Over two fifths (44%) of the participants benefited by getting fresh vegetables; over one third (35%) felt that their diet was improved by fresh vegetables they harvested, and over one fourth (26%) enjoyed gardening for personal satisfaction.

**Table 2.** Socioeconomic benefits of gardening.

| Benefit | | Percent |
|---|---|---|
| A. | Life quality | |
| | Fresh food/vegetables | 44 |
| | Improved diet | 35 |
| | Personal satisfaction and enjoyment | 26 |
| B. | Economic well-being | |
| | Saved money | 34 |
| C. | Social well-being | |
| | Socializing | 31 |
| | Helping others | 29 |
| | Sharing the produce with others | 15 |
| | Feeling of self-sufficiency | 14 |
| | Improved neighborhood | 13 |

### Economic Well-Being

An important economic "fringe benefit" for one third (34%) of the gardeners was "money saved." In 1989, there were 905 community gardens covering an area of about 15 acres, growing 45 varieties of vegetables worth over $450,000. The average size of a community garden was 720 square feet. This amount of space typically yields about one pound of vegetables per square foot. According to this formula adopted by the USDA

dollar value of production = area (sq. ft) × crop intensity × crop quality × length of season

for converting garden area into dollar value of production, the typical dollar yield per garden was over $500. The average supply cost per garden was about $25, making the average garden savings in the $475 range. The percentage of return on direct-dollar involvement is definitely enviable. This $475 saving is tax-free. Greater yields and dollar savings can be coaxed from the garden, depending upon the size of the plot, length of the growing season, and techniques used. Typical comments were, "I have hardly bought any vegetables since gardening." "I garden mainly to save money and provide vegetables to meet our family's needs year-round." "I plant varieties that I can't get at local markets or ones that are too costly." "My harvest is fresh and doesn't cost me anything."

The estimated cost to Rutgers Cooperative Extension for the Urban Gardening Program is $240,000 annually. When one considers the return value of vegetables produced each year (over $450,000) from vacant city lots formerly used as dumping grounds for junk cars and other trash, the program is cost-effective.

## Social Well-Being

The figures in Table 2 strongly indicate that the gardens became places for social interaction and community building. Almost one third (31%) of the gardeners developed new friendships through the gardens. Gardening promoted a community atmosphere and gave people an opportunity to meet others, share concerns, and solve a few problems together. Over one fourth (29%) helped others, and 15% shared their produce. Many gardeners commented: "Most of our gardeners enjoy the social experience. With many people here, lots of them might never meet if it weren't for the garden." "We have developed new friendships through gardening. We didn't know many people in our neighborhood until we started telling one another about how tasty our vegetables were."

About one eighth (13%) of the participants expressed that the gardening activity improved the neighborhood. Typical comments included: "It improved the neighborhood." "All vacant lots should be used." "It is better to have a garden instead of having a garbage-filled lot." "Even people just passing felt like stopping and talking to gardeners." "Over the garden, we know who our neighbors are." What stands out in this array of responses is that through gardening, participants felt good about themselves and their ability to cope with the world around them. It served as a neighborhood-building activity. People's behavior as a social group is modified by the presence of plants and participation in gardening activities. Gardening served as a way to break down some of the social barriers that existed between neighbors.

Perhaps the most significant benefit of community gardening for those who own no land is providing a piece of land for people to call their own.It is estimated that over 20% of U.S. land is held by corporations, much of it around cities and suburbs, where the need for gardening space is acute (Sommers, 1984). For homeless Americans, community gardening can be the first step toward self-sufficiency—providing land to garden, a place to call "mine," and the opportunity to grow and produce things of value for oneself.

## SUMMARY

In Newark, community gardening is the foundation for community development. Community gardening builds and rejuvenates neighborhoods and develops human capital.

It improves public relations. It brings people together in a context that enriches them and leads to positive interaction. It proved to be a major factor in contributing to community pride, thereby leading to improved litter control. For the neediest members of society, community gardening provided more than money, nutrition, and open space. It provided a sense of self-pride and self-worth in producing food on their own and for themselves. Community gardening is an activity that sells itself. The implications of community gardening are unlimited.

## LITERATURE CITED

Green, P. S. 1981. Gardeners' perceptions of the Newark urban gardening program: A program evaluation. Rutgers Cooperative Extension Report.

Program Staff. 1989a. Urban gardening program evaluation. Rutgers Cooperative Extension Report.

Program Staff. 1989b. Rutgers Cooperative Extension urban gardening 1989 annual report. Rutgers Cooperative Extension Report.

Sommers, L. 1984. Theory G—the employee gardening book. The National Association for Gardening, Burlington, Vermont.

# Community Gardening:
# A Model of Integration and Well-Being

James W. Reuter

Executive Director, Program and Family Support, The Bancroft School


Catherine M. Reuter, HTM

The Bancroft School

## ABSTRACT

In 1986, The Bancroft School, an organization serving adults and children with developmental disabilities, joined in a partnership with the borough of Haddonfield, New Jersey and the local garden club to create a community gardening project entitled "Lantern Lane." The impetus for the project was threefold: (1) the borough of Haddonfield wished to beautify its business district, and the area known as Lantern Lane stood out as an eyesore, (2) the Garden Club of Haddonfield wished to participate in borough beautification but needed personnel and/or technical assistance, and (3) Bancroft, which has had a well-developed horticulture therapy department for many years, wished to focus on community integration and community service, particularly with its population of older adults. From the outset, the project seemed to meet the needs and desires of all three organizations. The borough would make progress on the beautification of its business district, particularly Lantern Lane; the garden club would fulfill its commitment to the borough to maintain the mountain community gardens; and Bancroft would help the people it served to become further integrated into the community.

As staff members at Bancroft, we were both particularly excited about the chance for the people we serve to participate in a project in the center of the community where they would be highly visible, where they would be identified as being involved in the community, and where they would be viewed as people with "abilities." Our hope was that their involvement would be prized by the community, thereby enhancing their personal self-esteem.

On a philosophical level, we felt that it was important that individuals with developmental disabilities be seen and recognized as "givers" and not just "receivers" of services. This project would be a vehicle for demonstrating to those we serve how to give of themselves, their time and talents, to others and to their community, while experiencing the inherent satisfaction that comes from giving. It would also demonstrate to local citizens ways in which individuals who have developmental disabilities can give of their time and talents to enhance the lives of their neighbors and their community. The medium for this education was horticulture, and all the interactions between people and plants was horticulture therapy.

# Community Gardening as Job Training: Economic Impact

Chris Cordts

Coordinator, Urban Gardening Program, Denver Cooperative Extension, Colorado State University

## ABSTRACT

The Green Industry adds millions of dollars to the Colorado economy annually, but suffers from a lack of trained entry-level workers. Research is needed to determine if the education and training currently available in horticulture and landscaping meet the entry-level standards of the industry. In particular, secondary and community college curricula must be analyzed to determine commonalities, and to translate course content to competency measures. These competencies in turn must be referenced to those segments of the industry currently testing entry-level employees to certify skill levels. The transition from the education and training programs to the labor force has not received the attention necessary to address the shortage of trained entry-level Green Industry workers. Learning styles of participants in horticulture/landscaping vocational programs and publicly funded training programs need to be identified as these styles affect participation. With a competency-based curriculum, accurately tied to learning style and industry expectations, education and training programs in horticulture and landscaping can supply a steady source of entry-level Green Industry employees. Education and training outcomes can be measured as they impact the labor force.

# Identifying Research Needs for Urban Forestry in Quito, Ecuador

Deborah B. Hill

Department of Forestry, University of Kentucky

## ABSTRACT

A change in the city government of Quito in 1988 brought in a mayor who promoted urban forestry and natural resources. As a result, Fundacion Natura (FN), an indigenous environmental group, the U.S. Peace Corps (PC), and (Partners in the Americas (NAPA), a U.S. grassroots exchange program, worked with municipal workers on a program for "arborizing" Quito. Major problems include a lack of resources (financial/equipment/personnel), a lack of appropriate knowledge and training, and a lack of public awareness of the importance of urban trees.

Currently, there is little or no baseline data concerning the ecological, economic, and sociological/health aspects of an urban forest. Fundacion Natura just completed a tree list of potential species for urban forestry, but there is no tree inventory. Laws set aside 10% of every neighborhood for "greenspace" but do not specify whether for parks, street trees, or both. My NAPA work with FN and PC taught me how little urban forestry was based on research, as well as how important trees were for landslide prevention on Quito's highly erodible volcanic soils and for the availability of potable water, especially in marginal areas.

Research is needed on the ecological/sociological impacts of tree planting and the economics of tree management in the existing urban "forest" (maintenance vs. removal/replacement). Comparative ecological and socioeconomic studies on treed and treeless neighborhoods would be particularly useful for encouraging financial commitment to the continuing arborization of Quito.

# SECTION III

# PLANTS AND THE INDIVIDUAL

CHAPTER 15

# Influences of Passive Experiences with Plants on Individual Well-Being and Health

Roger S. Ulrich

Associate Dean for Research, College of Architecture, Texas A & M University

Russ Parsons

Research Associate, College of Architecture, Texas A & M University

## INTRODUCTION

About 80% of the American population lives in cities or urban areas. At the same time, most Americans would prefer to live in smaller towns or urban areas rather than in cities of 50,000 people or more (Louis Harris and Associates, 1978). This implies that for many people, the attractions of large cities, such as cultural amenities, shopping, and jobs, are outweighed by negative aspects such as crowding, crime, noise, long commutes, air pollution, and other stressors. For economic and other reasons, however, many people cannot live where they would like, so there is a clear need for changes in the environments of our cities that would make them more pleasant and liveable.

A widely held notion is that trees, flowers, and other vegetation help make cities better places for people. Unfortunately, the reality is that many of our highly urbanized areas contain little vegetation. Although some cities benefit from having moderate amounts of street trees and other urban forest amenities (Kielbaso, 1990), flowering plants and other smaller vegetation are often nearly nonexistent in public urban areas. Most of the people responsible for urban development—developers, politicians, and planners—probably agree that plants contribute to environmental quality. Relatively little scientific research about human-plant interactions, is available, however, and this can create the impression among decision-makers that there is an absence of tangible, credible evidence regarding the benefits that plants make possible. Unfortunately, intuitive arguments in favor of plants usually make little impression on financially pressed local or state governments or on developers concerned with the bottom line (Ulrich, 1980). Politicians, faced with urgent problems such as homelessness or drugs, may dismiss plants as unwarranted luxuries. The lack of research on plant benefits has also reduced spending for plants in other important settings, such as workplaces, health-care facilities, and outdoor areas of apartment complexes.

In recent years, however, researchers from several disciplines have begun investigating the benefits of contacts with plants, especially large-sized vegetation such as trees. The amount of research is still relatively small but is growing rapidly and has already deepened our understanding of the positive, need-satisfying experiences that plants make possible. Such studies, by contributing tangible, convincing evidence of the benefits of vegetation, may eventually help achieve higher priority for allocating funds to plants in indoor and outdoor environments where people live, work, learn, or receive health care. In particular, scientific research that documents the role of plants in fostering human physiological well-being will probably prove to be one effective means for gaining higher priority for plants.

People derive benefits from plants in a wide range of situations, including active contacts or involvement such as gardening, and through more physically passive experiences such as viewing flowers in a park or looking at plants through a window. In this paper we focus mainly on the influences of visual contacts with plants on psychological and physiological well-being and on health-related indicators. Particular emphasis is given to stress-reducing benefits of viewing vegetation. As becomes evident from the survey of research in later sections, most research findings to date pertain to large-sized vegetation such as trees or mixed vegetation dominated by trees and shrubs. Comparatively little research has been carried out to evaluate benefits derived from viewing small plants and flowers. In what follows, we first review theories from the social sciences that are relevant to explaining possible beneficial influences of visual contacts with vegetation. We then review findings from related fields, such as urban forestry and environmental psychology, that suggest that settings dominated by large vegetation such as trees have stress reducing and other beneficial effects. These findings bolster intuitions about possible benefits of flowers and other small plants in public and private settings. The review of findings relating to vegetation is divided into four parts: (1) aesthetic benefits; (2) effects on psychological well-being, including stress reduction; (3) physiological influences; and (4) health-related benefits. Finally we discuss promising directions for research on flowers and small plants, taking into account recent technological advances that create unprecedented opportunities for performing sound, scientific studies of the role of plants in fostering human well-being and health.

## BELIEFS AND THEORIES

The beliefs that contacts with trees, grass, and flowers foster psychological well-being and help reduce the stresses of urban living seem to be as old as urbanization itself. The villa gardens of the ancient Egyptian nobility and the walled gardens of Persian settlements in Mesopotamia indicate that the earliest urban peoples went to considerable lengths to maintain some direct contact with nature (Shepard, 1967). Numerous early writings suggest that vegetation and other nature were valued in cities. For instance, in the 1st Century B.C., the Roman poet Horace wrote regarding city dwellers: "Why, amid your varied columns you are nursing trees, and you praise the mansion which looks out on distant fields" (quoted in Glacken, 1967).

In the United States in the 1860s and 1870s, the renowned landscape architect Frederick Law Olmsted wrote at length about his intuitively based conviction that visual contact with nature, including plants, is beneficial to the emotional and physiological health of city dwellers. He asserted that an environment containing vegetation or other nature "employs the mind without fatigue and yet exercises it; tranquilizes it and yet enlivens it; and thus, through the influence of the mind over the body, gives the effect of refreshing rest and reinvigoration to the whole system" (1865). Olmsted's strong belief that vegetation and other nature in cities bring "tranquility and rest to the mind" was an important part of his eloquent justification for providing parks and vegetation in America's cities. Along with the famous parks he helped create, such as New York's Central Park, Olmsted's ideas about the healthful, restorative effects of nature in cities were a major influence on the City Beautiful movement and had widespread effects on parks and urban landscaping that continue to the present.

## CONTEMPORARY THEORIES

A century after Olmsted, authors from both the social and natural sciences have advanced a number of quite different theoretical perspectives that are relevant to explaining why people may derive enhanced well-being from passive contacts with flowers, trees, and other plants. Importantly, all these theoretical viewpoints, despite their differences, agree in predicting that passive experiences with environments having vegetation or other nature should tend to have positive effects on psychological and physiological well-being (Ulrich and Simons, 1986). A related point is that these theories all imply that contacts with environments having vegetation, compared to the effects of urban or built settings lacking nature, should usually be effective in fostering restoration from stress.

*Overload* and *arousal* theories have differences, but both propose that environments with high levels of visual complexity, noise, intensity, and movement can overwhelm and fatigue human perceptual systems, or lead to detrimentally high levels of psychological and physiological excitement (e.g., Cohen, 1978). Both theories imply that restoration from stress or perceptual fatigue should be fostered by settings having stimuli, such as plants, that are low in intensity and incongruity, and have patterning that reduces arousal and processing effort (e.g., Berlyne, 1971). There is some evidence that settings dominated by vegetation tend to have lower levels of complexity and other arousal-increasing properties than urban settings lacking nature (Wohlwill, 1976). Accordingly, these theories imply that surroundings containing prominent vegetation, compared to the effects of intense, perceptually jumbled urban settings, should have positive, stress-reducing effects on people.

Alternatively, another important category of theories emphasizes *learning* as the key mechanism for acquiring positive responses to plants and other nature. For instance, it might be argued that many individuals acquire positive associations with vegetation and other nature during vacations and other recreational experiences in rural areas. On the other hand, many people probably learn negative or stressful associations with urban environments because of such negative phenomena as crime or traffic congestion. An example relating more specifically to plants would be that learning and familiarity presumably explain why Arizona residents have more positive attitudes or responses to succulent shrubs and other desert plants than do Americans living in temperate parts of the country (e.g., Hecht, 1975). *Cultural* explanations also emphasize learning, suggesting that individuals are taught or conditioned by society to prefer certain environmental elements and dislike others or find them unsettling. A cultural argument could be used to explain, for instance, why the French may tend to like topiary, or why many Americans apparently prefer foundation plantings in their front yards. More generally, different authors have concluded that Western cultures condition their populations to like nature, including vegetation, and have negative associations with cities (e.g., Tuan, 1974). Hence cultural and other learning-based perspectives can suggest at least partial explanations for a given society's positive disposition to vegetation generally, and for greater liking for one particular plant variety over another.

Cultural and other learning-based perspectives are inadequate, however, for explaining why widely diverse cultures may believe in the restorative influences of nature. Such theories also show weaknesses in the face of growing empirical evidence indicating that there can be impressive similarity across Western and non-Western cultures in terms of greater liking for natural scenes in contrast to views of urban or built environments (e.g., Hull and Revell, 1989). Cultural explanations help to account for how responses or associations with respect to environmental elements are transmitted and maintained within a society, but they do not adequately explain, for instance, why the belief originated in a culture that contact with vegetation is restorative.

In recent years, authors have increasingly advanced *evolutionary* theoretical positions, partly because these perspectives are easily reconcilable with research findings of cross-cultural agreement in liking for vegetation and other nature. Further, evolutionary arguments contribute explanations for content-specific restorative influences and preferences (e.g., stress-reducing effects of vegetation as a general category of environmental content). Although evolutionary arguments can have pronounced differences, a premise shared by most authors is that the long evolutionary development of humankind in natural environments has left its mark on our species in the form of unlearned predispositions to pay attention and respond positively to certain contents (e.g., vegetation, water) and configurations that comprise those environments. People respond especially positively to combinations of contents and forms characteristic of natural settings that were most readily exploited by premodern humans, or were most favorable to ongoing well-being or survival. As an example, Orians (1986) has linked data indicating high aesthetic liking for certain vegetation and tree structures to scientific measurements showing a high potential for obtaining food and drinking water in such settings. Orians has also suggested that such evolutionary-based preferences underlie many "cultural" expressions of nature, such as gardens. In an interesting analysis, he has shown that empirical measurements of trees in high food–yielding savanna environments mirror closely the particular tree shapes selected and miniaturized for Japanese gardens. Another prominent evolutionary perspective has been advanced by the Kaplans (1982, 1989), who link cognitively based preferences and restorative influences to certain general contents (e.g., vegetation) as well as properties that facilitate exploring and making sense of settings (e.g., coherence). Ulrich (1983) has developed a "psychoevolutionary" perspective with the objective of explaining a broad range of emotional and physiological influences of natural configurations and content, including vegetation. Among other contrasts with the Kaplans' perspective, Ulrich postulates that quick-onset affective or emotional reactions—not cognitive responses—constitute the first level of response to nature, and are central to subsequent thoughts, memory, meaning, and behavior with respect to environments. This position is consistent with a large body of contemporary research on emotions and cognition and with recent advances in understanding neurophysiology (e.g., Öhman, 1986; LeDoux, 1986).

Regarding stress-reducing effects of vegetation, authors from different fields have postulated that strong attention-holding properties of nature may be an important mechanism in restoration. As a prominent example, Frederick Law Olmsted wrote insightfully about the mental stresses and fatigue associated with cities and "modern" life, and conjectured that natural views foster restoration because of their capacity to hold attention and block out the demands and distractions of daily work and living: "The attention is aroused and the mind occupied without purpose" (1865). The Kaplans have used William James' concept of "involuntary attention" in arguing that people respond with strong attention or "fascination" to nature, and this fosters restoration from mental fatigue associated with tasks or conditions that require sustained, disciplined, or taxing attention (Kaplan and Kaplan, 1989). Arguing from an evolutionary perspective, Katcher and his associates (Katcher et al., 1984) have suggested that a major reason why relaxation is induced by viewing a different type of natural configuration—an aquarium with fish—is that nature is effective in holding attention, diverting people's awareness away from themselves and from worrisome thoughts, and eliciting a meditation-like state. As is discussed below, Ulrich has performed a sequence of studies that have yielded direct empirical evidence suggesting that nature scenes dominated by vegetation, compared to urban scenes without vegetation, effectively hold attention and interest (e.g., Ulrich, 1979, 1981). These studies also indicate that along with attentional effects, vegetation elicits emotional and physiological responses that play critical roles in restoration.

To summarize briefly, the old belief that passive exposure to plants and other nature fosters human well-being is echoed in a number of contemporary theories that offer different explanations for positive responsiveness to nature. Although we consider evolutionary or unlearned factors to be of primary or most general importance, arousal/overload and learning-based explanations also have certain merits and can contribute to a more complete understanding of human responses to plants. Restorative and other beneficial effects of plants probably arise from a combination of factors and mechanisms, including evolutionary or unlearned influences, learning, and arousal-reducing properties of plants related to complexity and intensity levels that are nontaxing or require little processing effort.

## PSYCHOLOGICAL WELL-BEING

### Aesthetic Benefits

One comparatively narrow but important category of psychological benefits of plants, and the one that has received the most attention from researchers, is the aesthetic. If viewing a setting with plants elicits a response of aesthetic liking or preference, then presumably an individual's feeling state may be somewhat more positively toned (Ulrich, 1990). Consistent with predictions suggested by the various theoretical perspectives described above, many studies conducted in different countries have shown that people usually accord higher liking to nature scenes dominated by vegetation than to urban scenes lacking vegetation (e.g., Kaplan et al., 1972; Zube et al., 1975). Within American urban environments, the presence of trees, grass, and other plants in settings usually increases aesthetic liking. Similarly, research in Japan and Western Europe suggests that urban scenes having vegetation are more preferred than urban settings lacking vegetation (e.g., Asakawa, 1984). This finding has been reported for urban settings such as residential areas (e.g., Cooper-Marcus, 1982; Nasar, 1983; Schroeder and Cannon, 1983; Zoelling, 1981); commercial streets and strips (e.g., Lambe and Smardon, 1986), and parking lots (Anderson and Schroeder, 1983). Several investigators have found that parks or park-like features having trees and other prominent vegetation are often especially preferred visual amenities in urban areas (e.g., Ulrich and Addoms, 1981). By contrast, urbanites respond with moderately low liking to empty grass-covered spaces lacking trees and other vegetation. In parks, aesthetic liking is particularly high for well-maintained areas having visual openness, scattered trees, and understory plantings that do not obscure foreground surveillance or create enclosure. Dense understory vegetation that restricts visual openness reduces aesthetic liking sharply and elicits feelings of insecurity (Daniel and Boster, 1976; Schroeder and Anderson, 1984; Hull and Harvey, 1989). Compared to whites, black Americans respond with even lower liking to settings having dense foreground vegetation or with a sense of enclosure (R. Kaplan and Talbot, 1988). [For surveys of research on aesthetic responses to trees and other large vegetation in cities, see Schroeder (1989) and Smardon (1988)].

Although a substantial body of research has focused on the aesthetic influences of trees and other large vegetation in cities, there is a shortage of studies on aesthetic responses to small plants and flowers. One of the few studies to consider flowers has been performed by Schroeder and Lewis (Schroeder, 1986), who asked members and volunteers at the Morton Arboretum to give verbal descriptions of areas they remembered at the Arboretum. Of the individuals in the study, 90% provided comments indicating that attractive views containing flowers were positive memories of the Arboretum—e.g., "fields of daffodils in an area of large oaks."

## Stress-reducing Effects of Settings with Vegetation

A stress reaction is the process of responding psychologically, physiologically, and often with behaviors to a situation that is taxing or threatens well-being (Evans and Cohen, 1987). Although certain short-term stressful situations can improve human performance and cognitive functioning, stress is considered here to be a negative condition that should be mitigated over time to prevent deleterious effects on human performance, well-being, and health. Although preference or aesthetic liking is an important emotional response, it is only one component of the broad range of feelings (e.g., fear, anger, sadness, interest) that are central to the psychological dimension of stress and restoration (Ulrich, 1983).

Consistent with the beliefs of Olmsted and others, and with predictions of the theories surveyed above, a large body of research on recreational experiences has indicated that leisure activities in nature settings with vegetation are important for helping people cope with stress as well as in meeting other non-stress-related needs. Most of this research focuses on benefits derived from experiences in wilderness environments, but a growing number of studies have assessed the psychological effects of leisure experiences in urban parks, botanical gardens, and yards and common areas of housing developments. A consistent finding in more than 100 wilderness studies has been that psychological restoration through stress reduction is one of the most important verbally expressed, perceived benefits (e.g., Driver and Knopf, 1975; Knopf, 1987). Similarly, restoration from stress has emerged as a key perceived benefit in most of the research on urban parks and green spaces in residential areas (e.g., Davis, 1973; Ulrich and Addoms, 1981; R. Kaplan, 1983; Hayward and Weitzer, 1984; Talbot et al., 1987).

Apart from restoration derived through viewing vegetation and other nature, other mechanisms probably contribute to stress recovery in these studies, including factors such as physical exercise, achieving a "breather," and achieving a sense of control with respect to work pressures and other stressors through "temporary escape" (Driver and Knopf, 1975) or "being away" (R. Kaplan and Talbot, 1983). Nonetheless, part of the restoration derived from such recreation experiences apparently stems from viewing vegetation and other nature. This conclusion is supported by a few park studies that have found statistical associations between reported restoration and questionnaire items relating to a park's appearance—e.g., trees, grass, open space (Ulrich and Addoms, 1981). Also, some research has identified restorative effects while controlling for variables such as physical exercise and psychological "escape." Hartig, Mang, and Evans (1987) produced stress in subjects with a demanding cognitive task, and then measured recovery produced by either (1) reading magazines or listening to music for 40 minutes, (2) walking in an urban area for 40 minutes, or (3) walking for an equivalent period in a nature area dominated by trees and other vegetation. Findings showed that individuals who had taken the nature walk had more positively toned feelings than subjects assigned to the other conditions. Schroeder (1986) found that the most common feelings associated with visits to the Morton Arboretum were serenity, tranquility, or peacefulness. Such feelings were often linked to experiences with settings having lush vegetation and openness.

A study framed explicitly as a test of Olmsted's "tranquility hypothesis" has yielded direct evidence of the restorative effects of merely viewing vegetation. Ulrich (1979) studied two groups of university students who were experiencing mild stress because of a final course exam. One group was shown a collection of color slides of unblighted urban scenes lacking vegetation; the other group was exposed to slides of rural settings dominated by trees and other vegetation. Consistent with Olmsted's conjecture, findings obtained from self-ratings of feeling states suggested that the vegetation scenes fostered greater stress recovery as indicated by sharp increases in levels of positive feelings, significant reductions in fear, and somewhat lower anger/aggression. Ulrich also found that the settings with vegetation sustained attention more effectively than the urban scenes lacking vegetation. Honeyman (1987) replicated this study with an important change—she tested an additional recovery condition consisting of slides of urban scenes *with* vegetation. Her findings suggested that greater restoration was produced by the urban scenes with vegetation than the settings without vegetation.

Another line of research, on window views and windowless settings, has provided additional evidence suggesting that visual contact with vegetation and other nature can be preferred and restorative. Compared to settings with windows, windowless rooms tend to be disliked and can be stressful, especially in workplaces and health-care settings (e.g., Keep et al., 1980; Ruys, 1970). Heerwagen and Orians (1986) found that office workers with little or no visual access to the outside were more likely to decorate their work spaces with posters and other depictions of outdoor scenes than were workers with windows. Further, most of the outdoor pictures used by the windowless group displayed settings dominated by vegetation and other nature. The windowless workers may have displayed nature pictures to compensate for the stressful influences of windowlessness (Heerwagen, 1990). In interiors with windows, views having depth, vegetation, or other nature are preferred over low-depth and visually impoverished window views (Markus, 1967; Verderber, 1986).

Other evidence implying the important restorative effects of nature in stressful interior environments has emerged from interviews with astronauts and cosmonauts. Space vehicles and facilities are isolated, cramped, hazardous, stressful environments. Wise and his associates interviewed a culturally diverse group of Western astronauts and Soviet cosmonauts, asking for suggestions for interior decor options they would prefer (Wise & Rosenburg, 1988). As summarized by the researchers in their report, the responses indicated a strong, widely shared preference for having more plants and other nature elements in stressful orbital environments:

> Respondents were nearly unanimous in asking for more natural and varied colors, plants, landscape pictures, and natural woods, regardless of their particular national origin. Human beings' love for nature and natural materials and forms, especially in high technology habitats, seems to transcend national boundaries. The Soviets' extensive use of natural scenes and working gardens to maintain morale in their Salyut and Mir crews is already well documented (Wise et al., 1990).

## PHYSIOLOGICAL EVIDENCE OF BENEFITS

In addition to psychological manifestations, stress and restoration have very important physiological dimensions. The physiological component is reflected in responses or levels of activity in numerous bodily systems, such as the cardiovascular. Data obtained by recording physiological responses are widely recognized to have scientific credibility as indicators of stress and restoration. Also, physiological methods can identify influences on well-being that may be outside the conscious awareness of individuals and hence may not be identified by verbal measures such as ratings or questionnaires.

In a study performed in Sweden (Ulrich, 1981), brain electrical activity was recorded from unstressed individuals while they viewed lengthy slide presentations of outdoor scenes. The major finding was that alpha wave activity was higher when subjects viewed nature settings dominated by vegetation as opposed to urban scenes lacking vegetation. Apart from indicating that the nature and urban scenes had different effects on electrocortical activity, the alpha wave results strongly suggested that the vegetation views were more effective in eliciting a wakeful, relaxed state. In the same study, self-ratings data suggested that the vegetation settings sustained attention/interest at higher levels than did the urban scenes, and produced more positively toned emotional states.

Physiological measures have also been used to study directly the question of stress-reducing effects of visual experiences with nature. Ulrich and Simons (1986) monitored a battery of physiological responses while stressed subjects experienced a "recovery" period consisting of 10-minute color/sound videotapes of either natural or urban outdoor environments. Results indicated that people recovered more quickly and completely from stress when exposed to the natural settings, which included a park-like setting dominated by vegetation. Greater recovery during the nature exposures was indicated by lower blood pressure, muscle tension, and skin conductance. The heart-rate response was of particular interest in this study. Heart-rate data suggested that the natural environments elicited considerable attention or perceptual intake, whereas the urban settings were perceptually rejected. The nature settings also fostered more recovery in the psychological component of stress as suggested by greater reductions in self-rated fear and anger, and much greater increases in positive feelings. The physiological findings indicated that the nature settings produced significant recovery from stress in only 4–6 minutes. This rapid recovery raises the possibility that comparatively brief visual contacts with vegetation might be important for many city dwellers from the standpoint of fostering restoration from mild daily stressors such as commuting and work pressures.

Recently, researchers have begun to use physiological measures to investigate stress-reducing effects of nature scenes in certain health-care and workplace settings. Heerwagen and Orians studied stressed patients in the waiting room of a dental fears clinic (Heerwagen, 1990). On some days, the researchers hung a large mural on a waiting room wall depicting a view of distant mountains, clustered trees, and open grassy areas. On other days, the wall was blank. Findings obtained from self-ratings of feelings suggested that patients felt calmer or less stressed on days when the scene with vegetation and other nature was on the wall. Likewise, heart-rate measurements also indicated that individuals were less stressed or tense when the nature scene was visible.

Wise and Rosenberg (1988) studied the role of nature decor in alleviating physiological stress in the context of work productivity of astronauts in a space station. Subjects were studied individually as they performed a series of stressful tasks in a simulated space station crew cabin at a NASA center. Each subject was exposed to one of four different pictures that was affixed to a cabin bulkhead: savanna-like nature; mountain waterfall; "hi-tech" abstract; and no picture (control). Although the mountain water scene was most aesthetically preferred, physiological data (skin conductance) suggested that the savanna-like scene was significantly more effective than any of the other visual conditions in mitigating stress. A most interesting finding was that the presence of the savanna scene apparently reduced stress even when subjects were not looking at it, or perhaps were not consciously aware of the scene. To account for this effect, Wise and Rosenberg speculated that the savanna scene elicited a positive affective state that provided a persistent buffer against stress during task performance. Similarly, findings from the dental fear clinic study described above raised the possibility that patients did not have to be looking at the nature scene, and perhaps not even be consciously aware of its presence, to derive restorative benefits (Heerwagen, 1990).

These physiological findings justify the speculation that people may not have to be consciously aware of the presence of plants in homes, workplaces, or other settings for the plants to have positive influences on emotional states and physiological indicators. Another implication of these physiological studies is that research approaches based on verbal ratings or evaluations of physical settings having plants (e.g., satisfaction or pleasantness ratings of a setting) may sometimes not reveal the effects of plants on well-being.

## HEALTH-RELATED BENEFITS

The findings surveyed in the preceding sections suggest that short-term exposures to vegetation can be effective in fostering recovery from mild stress. Accordingly, it seems possible that the potential benefits of viewing trees, flowers, and other vegetation may often be greatest when individuals experience considerable stress or anxiety and are required to spend long periods in a confined setting (Ulrich, 1979, 1981). Such situations include, for instance, health-care contexts, prisons, and certain high-stress work environments. In these and other settings, long-term exposures to views of vegetation and other nature may have persistent positive influences on psychological and physiological well-being, functioning, and possibly behaviors—influences that might in turn be reflected in health-related indicators.

Findings from a few studies focusing on prisons and hospitals suggest that window views of vegetation or other nature can have important health-related benefits. Moore (1982) examined the need for prison health-care services by inmates whose cells looked out onto the prison yard and those who had a view of nearby farmlands and forests. He reported that inmates who had the natural view were less likely to report for sick call. Likewise, West (1985) found that cell window views of nature, compared to views of prison walls, buildings, or other prisoners, were associated with lower frequencies of health-related stress symptoms such as headaches and digestive upsets. Ulrich (1984) compared the hospital records of matched pairs of gall bladder surgery patients who had window views of either a small stand of trees or a brick building wall. He found that patients with the view of trees had shorter post-operative hospital stays, required fewer potent pain drugs, and received fewer negative staff evaluations about their conditions than those with the wall view.

Studies such as these using health-related measures suggest opportunities for linking economic benefits to passive experiences with vegetation and other nature. For instance, because patients in the hospital study required fewer costly pain injections, and prisoners needed fewer health services, it seems likely there are dollar savings associated with the views of nature.

## RESEARCH DIRECTIONS: ADVANCES IN METHODS AND EQUIPMENT

There have been major advances recently in the development of research equipment and new techniques that have created unprecedented opportunities for performing sound, quality research on the beneficial influences of contacts with plants. These advances will enable researchers, for instance, to apply physiological and health-related procedures in a broad range of real-world and laboratory situations—e.g., home interiors, backyard gardens, workplaces, health-care facilities, botanical gardens, shopping malls, and urban settings. Achieving this potential will require increased availability of research funding, however, and the bringing together of multidisciplinary research teams having the necessary range of expertise.

Many of the new opportunities stem from advances in electronics miniaturization and computers that are making the recording of important physiological indicators (e.g., blood pressure, heart rate) increasingly practical for investigating the effects of plants on well-being. Compact, self-contained unobtrusive units that record or transmit data can be worn by individuals as they experience settings having flowers or other plants. Eye-tracking equipment has become available that is comparatively inexpensive and unobtrusive, and can be worn by individuals, for instance, while they walk through gardens or drive through urban areas. The eye-tracking apparatus makes it possible to study such issues as the extent to which people notice and pay attention to flowers and other plants. This information could have considerable practical value in guiding decisions, for example, about where to locate ornamental plantings in parks or pedestrian shopping areas so that the visual impact is maximized. Eye-tracking studies would also be very useful in generating guidelines regarding size and site requirements for roadside plantings to ensure that fast-moving motorists notice and benefit from the plants.

There has also been rapid progress in developing techniques for simulating environments for research. For instance, the Visualization Laboratory at Texas A&M University has the capability to present realistically existing or imagined environments via computer graphics and animation. Settings can be displayed with a realism potentially equivalent to photographic quality in three-dimensional color graphics that may be either static or animated. This technology enables researchers, for instance, to study individuals' reactions to scenes that vary systematically with respect to the presence of plant species or flower colors. Animation techniques make it possible for an observer at a large monitor to "walk through" a garden or shopping area, or "drive through" an urban area before and after plants or other landscaping are added. These kinds of advances, both in environmental simulation and in techniques for scientifically measuring effects on well-being, should open doors to significant progress in understanding the human benefits of plants.

## SUMMARY AND CONCLUSIONS

The old belief is that visual contact with plants and other nature is somehow good for people, and can help individuals cope with the stresses of urban living. Contemporary theories, whether they emphasize learning or evolutionary explanations, agree in predicting that visual contacts with environments having vegetation or other nature should tend to have positive effects on psychological and physiological well-being. Progress has been made in investigating the benefits derived from visual experiences with trees and other large vegetation; however, little work has focused on the role of views of flowers and small plants in fostering psychological and physiological well-being. Relatively little is known in a scientific sense about issues such as aesthetic preferences for different small plants, effects of viewing small plants on emotional states and stress recovery, physiological influences of viewing flowers, or possible health-related effects associated with long duration or frequent exposures to flowers and other plants.

A large body of research has shown that the presence of trees and other large vegetation in urban settings enhances aesthetic liking or preference. Also, a growing number of studies have found that viewing nature scenes dominated by vegetation has beneficial effects on psychological and physiological well-being, and in certain situations can have positive effects on health-related indicators. In laboratory research, visual exposure to settings with vegetation has produced significant recovery from stress within only five minutes, as indicated by changes in physiological measures such as blood pressure and muscle tension. Views of vegetation foster restoration from stress apparently because of a combination of beneficial effects: They produce increases in positive feelings; reduce negatively toned or stress-related feelings such as fear, anger, or sadness; hold interest/attention effectively and hence may block or reduce stressful thoughts; and elicit positive changes across different physiological systems. This combination of positive psychological and physiological effects observed for short-duration exposures may underlie beneficial health-related influences of vegetation views found in studies of stressful real-world settings such as hospitals and prisons. Vegetation views that are comparatively effective in reducing stress may not always be those that rate highly in aesthetic preference. Although direct visual attention is important in restoration, recent findings raise the possibility that stress reduction may persist without attention, and perhaps not even require conscious awareness that plants or views of vegetation are in the individual's immediate surroundings. In any case, it seems clear that the benefit of viewing vegetation goes far beyond aesthetics to include a range of other effects important to well-being.

Among other research needs, advances in understanding the benefits of plants will require many more studies that assess effects on well-being by using physiological and health-related measures. The advantages of this research direction are being enhanced by rapid advances in electronics miniaturization and computers that make it possible for researchers to use physiological procedures more easily in real-world settings (e.g., gardens, workplaces, health-care facilities) and in laboratory situations. Research on the benefits of plants that uses physiological and health-related methods offers important advantages. For instance, research findings obtained from physiological procedures (e.g., blood pressure effects) tend to have scientific credibility, often carry weight in decision-making and in environmental impact statements, and can be considered permissible data by courts, where such findings can sometimes be reconciled with legal interpretations of public health and welfare. If researchers wish to study the effects of plants on well-being or health, methods or measures should be used that directly relate to well-being and health, including physiological and other methods used in such fields as health psychology, clinical psychology, and behavioral medicine. By contributing tangible, convincing evidence of the importance of plants for human well-being and health, future research that uses physiological and health-related methods will likely help horticulturists gain higher priority for plants in spending or allocation decisions.

## ACKNOWLEDGMENTS

Portions of the research discussed in this paper were supported by USDA-Forest Service Co-operative Agreements Nos. 28-C7-420 and 28-C7-424, with the Rocky Mountain Forest and Range Experiment Station. Other support was provided by National Science Foundation grant SES-8317803.

## LITERATURE CITED

Anderson, L. M. and H. W. Schroeder. 1983. Applications of wildland scenic assessment methods to the urban landscape. Landscape Planning 10:219–237.

Asakawa, S. 1984. The effects of greenery on the feelings of residents towards neighborhoods. Journal of the Faculty of Agriculture, Hokkaido University 62:83–97.

Berlyne, D. E. 1971. Aesthetics and psychobiology. Appleton-Century-Crofts, New York.

Cohen, S. 1978. Environmental load and the allocation of attention. In: A. Baum, J. E. Singer, and S. Valins (eds.). Advances in environmental psychology. vol. 1. Lawrence Erlbaum, Hillsdale, New Jersey.

Cooper-Marcus, C. 1982. The aesthetics of family housing: The residents' view-point. Landscape Research 7:9–13.

Daniel, T. C. and R. S. Boster. 1976. Measuring landscape esthetics: The scenic beauty estimation method. USDA Forest Service Research Paper RM–167, 66 pp.

Davis, R. L. 1973. Selected motivational determinants of the recreational use of Belle Isle Park in Detroit. Unpublished Master's Thesis. School of Natural Resources, University of Michigan, Ann Arbor.

Driver, B. L. and R. C. Knopf. 1975. Temporary escape: One product of sport fisheries management. Fisheries 1(2):24–29.

Evans, G. W. and S. Cohen. 1987. Environmental stress, p. 571–610. In: D. Stokols and I. Altman (eds.). Handbook of environmental psychology. John Wiley, New York.

Glacken, C. J. 1967. Traces on the Rhodian shore: Nature and culture in western thought from ancient times to the end of the eighteenth century. Univ. of Calif. Press, Berkeley.

Hartig, T., M. Mang, and G. W. Evans. 1987. Perspectives on wilderness: Testing the theory of restorative environments. Paper presented at the Fourth World Wilderness Congress, Estes Park, Colorado.

Hayward, D. G. and W. H. Weitzer. 1984. The public's image of urban parks: Past amenity, present ambivalence, uncertain future. Urban Ecology 8:243–268.

Hecht, M. E. 1975. The decline of the grass lawn tradition in Tucson. Landscape 19:3–10.

Heerwagen, J. H. 1990. The psychological aspects of windows and window design. In: R. I. Selby, K. H. Anthony, J. Choi and B. Orland (eds.). Proceedings of 21st Annual Conference of the Environmental Design Research Association, Champaign-Urbana, Illinois, 6–9 April.

Heerwagen, J. H. and G. Orians. 1986. Adaptations to windowlessness: A study of the use of visual decor in windowed and windowless offices. Environment and Behavior 18:623–639.

Honeyman, M. 1987. Vegetation and stress: A comparison study of varying amounts of vegetation in countryside and urban scenes. Unpublished Master's Thesis. Department of Landscape Architecture, Kansas State University, Manhattan.

Hull, R. B. and A. Harvey. 1989. Explaining the emotion people experience in suburban parks. Environment and Behavior 21:323–345.

Hull, R. B. and G. R. B. Revell. 1989. Cross-cultural comparison of landscape scenic beauty evaluations: A case study in Bali. Journal of Environmental Psychology 9:177–191.

Kaplan, R. 1983. The role of nature in the human context. In: I. Altman and J. J. Wohlwill (eds.). Human behavior and environment, Vol. 6, Behavior and the natural environment. Plenum Press, New York.

Kaplan, R. and S. Kaplan. 1989. The experience of nature. Cambridge, New York.

Kaplan, R. and J. F. Talbot. 1988. Ethnicity and preference for natural settings: A review and recent findings. Landscape and Urban Planning 15:107–117.

Kaplan, S. and R. Kaplan. 1982. Cognition and environment. Praeger, New York.

Kaplan, S., R. Kaplan and J. S. Wendt. 1972. Rated preference and complexity for natural and urban visual material. Perceptual Psychophysics 12:354–356.

Kaplan, S., and J. F. Talbot. 1983. Psychological benefits of a wilderness experience. In: I. Altman and J. F. Wohlwill (eds.). Human behavior and environment, Vol. 6, Behavior and the natural environment. Plenum Press, New York.

Katcher, A., H. Segal, and A. Beck. 1984. Comparison of contemplation and hypnosis for the reduction of anxiety and discomfort during dental surgery. American Journal of Clinical Hypnosis 27:14–21.

Keep, P., J. James, and M. Inman. 1980. Windows in the intensive therapy unit. Anaesthesia 35:257–262.

Kielbaso, J. J. and V. Cotrone. 1990. The state of the urban forest. In: P. D. Rodbell (ed.). Proceedings of the Fourth Urban Forestry Conference. American Forestry Association, Washington, D.C.

Knopf, R. C. 1987. Human behavior, cognition and affect in the natural environment. p. 783–825. In: D. Stokols and I. Altman (eds.). Handbook of environmental psychology. John Wiley, New York.

Lambe, R. A., and R. C. Smardon, 1986. Commercial highway landscape reclamation: A participatory approach. Landscape Planning 12:353–385.

LeDoux, J. E. 1986. Sensory systems and emotions: A model of affective processing. Integrative Psychiatry 4:237–248.

Louis Harris and Associates, Inc. 1978. The 1978 HUD survey of the quality of community life, a data book. U.S. Government Printing Office, Washington, D.C.

Markus, T. A. 1967. The function of windows: A reappraisal. Building Science 2:97–121.

Moore, E. O. 1982. A prison environment's effect on health care service demands. Journal of Environmental Systems. 11(1):17–34.

Nasar, J. L. 1983. Adult viewers' preferences in residential scenes: A study of the relationship of environmental attributes to preference. Environment and Behavior 15:589–614.

Öhman, A. 1986. Face the beast and fear the face: Animal and social fears as prototypes for evolutionary analyses of emotion. Presidential address given at the 1985 Meeting of the Society for Psychophysiological Research. Psychophysiology 23(2):123–145.

Olmsted, F. L. 1865. The value and care of parks. Report to the Congress of the State of California. (Reprinted in Landscape Architecture 17:20–23, 1952, and in Nash, R. 1976. The American environment. Addison-Wesley, Reading, Mass.).

Orians, G. H. 1986. An ecological and evolutionary approach to landscape aesthetics, pp. 3–25. In: E. C. Penning-Rowsell and D. Lowenthal (eds.). Meanings and values in landscape. Allen and Unwin, London.

Ruys, T. 1970. Windowless offices. Master's Thesis, College of Architecture, University of Washington, Seattle. Cited in Heerwagen and Orians, 1986.

Schroeder, H. W. 1986. Psychological value of urban trees: Measurement, meaning and imagination. Proceedings of the Third National Urban Forestry Conference. American Forestry Association, Washington, D.C.

Schroeder, H. W. 1989. Environment, behavior, and design research on urban forests. p. 87–117. In: E. H. Zube and G. T. Moore (eds.). Advances in environment, behavior and design. vol. 2. Plenum, New York.

Schroeder, H. W. and L. M. Anderson. 1984. Perception of personal safety in urban recreation sites. Journal of Leisure Research 16:177–194.

Schroeder, H.W. and W. N. Cannon, Jr. 1983. The esthetic contribution of trees to residential streets in Ohio towns. Journal of Arboriculture 9:237–243.

Shepard, P. 1967. Man in the landscape: A historic view of the esthetics of nature. Alfred A. Knopf, New York.

Smardon, R. C. 1988. Perception and aesthetics of the urban environment: Review of the role of vegetation. Landscape and Urban Planning 15:85–106.

Talbot, J. F., L. V. Bardwell, and R. Kaplan. 1987. The functions of urban nature: Uses and values of different types of urban nature settings. Journal of Architectural and Planning Research 4:47–63.

Tuan, Y. F. 1974. Topophilia: A study of environmental perception, attitudes and values. Prentice Hall, Englewood Cliffs, New Jersey.

Ulrich, R. S. 1979. Visual landscapes and psychological well-being. Landscape Research 4(1):17–23.

Ulrich, R. S. 1980. Benefits of passive experiences with plants. Longwood Program Seminars. University of Delaware and Longwood Gardens, Newark, Delaware.

Ulrich, R. S. 1981. Natural versus urban scenes: Some psychophysiological effects. Environment and Behavior 13:523–556.

Ulrich, R. S. 1983. Aesthetic and affective response to natural environment, pp. 85–125. In: I. Altman and J. F. Wohlwill (eds.). Human behavior and environment. vol. 6. Plenum, New York.

Ulrich, R. S. 1984. View through a window may influence recovery from surgery. Science 224:420–421.

Ulrich, R. S. 1990. The role of trees in human well-being and health, pp. 25–30. In: P. D. Rodbell (ed.). Proceedings of the Fourth Urban Forestry Conference. American Forestry Association, Washington, D.C.

Ulrich, R. S. and D. L. Addoms. 1981. Psychological and recreational benefits of a residential park. Journal of Leisure Research 13:43–65.

Ulrich, R. S. and R. F. Simons. 1986. Recovery from stress during exposure to everyday outdoor environments. In: J. Wineman, R. Barnes, and C. Zimring (eds.). The Costs of Not Knowing: Proceedings of the Seventeenth Annual Conference of the Environmental Design Research Association. Environmental Design Research Association, Washington, D.C.

Verderber, S. 1986. Dimensions of person-window transactions in the hospital environment. Environment and Behavior 18:450–466.

West, M. J. 1985. Landscape views and stress response in the prison environment. Unpublished Master's Thesis. Department of Landscape Architecture, University of Washington, Seattle.

Wise, J. A., K. McConville, and N. Al-Sahhaf. 1990. Managing cultural diversity in orbiting habitats. In: Proceedings of Space '90, Albuquerque, New Mexico.

Wise, J. A. and E. Rosenberg. 1988. The effects of interior treatments on performance stress in three types of mental tasks. CIFR Technical Report No. 002-02-1988. Grand Valley State University, Grand Rapids, Michigan.

Wohlwill, J. F. 1976. Environmental aesthetics: The environment as a source of affect, pp. 37–86. In: I. Altman and J. Wohlwill (eds.). Human behavior and environment. vol. 1. Plenum, New York.

Zoelling, M. M. 1981. Urban high-rise dwellers' visual perceptions of form and space in the central business district of Syracuse, New York. Unpublished Master's Thesis, Landscape Architecture, SUNY ESF, Syracuse, NY.

Zube, E. H., D. G. Pitt, and T. W. Anderson. 1975. Perception and prediction of scenic resource values of the Northeast, pp. 151–167. In: E. H. Zube, R. O. Brush, and J. G. Fabos (eds.). Landscape assessment: values, perceptions and resources. Dowden, Hutchinson & Ross, Stroudsburg, Pennsylvania.

CHAPTER 16

# Effects of Plantscapes in an Office Environment on Worker Satisfaction

Kim Randall,* Candice A. Shoemaker,** Diane Relf, † E. Scott Geller, ‡

*Graduate Student, Department of Psychology, **Research Associate of Horticulture, † Associate Professor of Horticulture, ‡ Professor of Psychology, Virginia Polytechnic Institute and State University

## INTRODUCTION

Survey research has shown that men and women generally report a preference for the presence of plants in indoor and outdoor settings. Rachel Kaplan (1985) has done extensive research on the preferences for natural areas and trees in the outdoor environment and in 1985 showed a link between the presence of trees/natural areas and neighborhood satisfaction. Kalmbach and Kielbaso (1979) have also shown that a natural outdoor environment is preferred to a man-made environment.

Large companies incorporate interior plantscaping as part of their policy. This practice is not founded on empirical research, although the companies claim that the plants increase productivity, decrease absenteeism, and improve morale (Horsbrough, 1972; Lewis, 1972; and Kaplan, 1978). In other research, Laviana (1985) has shown that plants have an effect on the perceived quality of indoor space and the human affective state.

When the opportunity became available to study the link between job satisfaction and plants in the interior office design in co-operation with researchers on a separate air quality study, we hypothesized that the presence of plants in the work place would increase the employee's positive self-report of environmental quality, worker attitudes, and job satisfaction.

The site for the research was chosen based on the requirements to conduct the air quality study; that is, separate air systems for each floor of a single company. The building used in the study is located in northern Virginia and consists of eleven floors, two of which (the ninth and eleventh floors) contain the company under study. The company works on complex computer system analysis. The design of the ninth floor is mainly a traditional office arrangement of enclosed work spaces. The eleventh floor consists of open-plan office space.

By using a single office complex on two levels, we have established a "within subjects design," which is especially useful in this type of research. It enables the researcher to determine the before-and-after effect of the implementation of plantscapes. In addition, this design eliminates the threat of complications such as significant effects due to lapse of time or overall change in the environment. For these reasons, a "within subjects design" provides tighter control of the experimental setting for scientific research.

The research started with the removal of all personal plants from both floors for three months to enable the air quality study to obtain baseline measurements. All employees were informed of the indoor air quality research project and instructed not to have any plant material in their office spaces until after their floor was plantscaped. The first questionnaire was administered to all employees three months after removal of the plants, and one week before the eleventh floor was professionally plantscaped. The employees on the ninth floor were reminded not to have plant material in their offices. At the same time plants were installed on the eleventh floor, artwork was hung on the ninth floor so that both floors received environmental changes.

The second questionnaire was administered six months after the installation of plants on the ninth floor. (The plants were maintained by a professional interior plant maintenance company.)

The first survey consisted of 41 questions; 25 of these questions were answered on a five-point, Likert-type scale (i.e. 1 = "strongly disagree"...... 5 = "strongly agree"). The remaining 16 questions were either multiple choice or fill-in-the-blanks. Employee code numbers were requested to track answers across different questionnaire administrations. All employees were informed that their answers would be confidential and at no time would the company examine individual answers. Eight of these questions pertained to the perceived quality of work space. Two of these questions determined an individual's desire for control over plants in a personal office or the placement of artwork in a personal office. The remaining 15 questions asked about the employees' current opinions of plants and artwork in various settings and the role of plants in their daily lives. Demographic questions assessed the respondents' work location, gender, and number of cigarettes smoked per day. Other questions asked about their knowledge of the research project and activities in which they participated.

This study is still in progress and the final results have yet to be collected and analyzed, though preliminary results show the workers' attitudes are favorable towards the plants (Table 1) and their work space (Table 2). In addition, there were only a few significant changes in workers' attitudes between the two questionnaires (Table 3). These results suggest the following:

1. A ceiling effect may be likely, since the scores were relatively high before the plants were introduced to the office and showed little change after the eleventh floor was plantscaped. It appears that the employees are apparently quite satisfied with their work space and have good opinions about plants.

2. The five-point scale may not be sensitive enough for our purposes.

Table 3 shows the differences that were found between the two groups. The employees stated more of a preference to have their own plants in their offices and to care for the plants themselves on the first questionnaire than on the second. This suggests that the employees were pleased with the professional plantscaping and maintenance on the eleventh floor and, therefore, did not feel it was as important to have and care for their own plants. In addition, on the second questionnaire, employees reported noticing the plants more than they did at

**Table 1.** Means: employee appreciation of plants in questionnaires #1 and #2[1].

|  | Questionnaire #1 | Questionnaire #2 |
|---|---|---|
| **Positively worded questions** | | |
| Being around plants makes me feel calmer and more relaxed. | 3.83 | 3.90 |
| Plants in the office make it a more desirable place to work. | 4.18 | 4.36 |
| Plants improve the air quality of an office. | 4.29 | 4.12 |
| **Negatively worded questions** | | |
| Plants in the office are not important beyond their aesthetic appeal. | 1.98 | 2.32 |

**Table 2.** Means: employee perception of office space in questionnaires #1 and #2[1].

|  | Questionnaire #1 | Questionnaire #2 |
|---|---|---|
| I like the space in which I work. | 3.96 | 3.76 |
| I have sufficient space to accomplish my work. | 3.63 | 3.70 |
| The noise level in my office is very good for the work I do. | 3.73 | 3.33 |
| The lighting in my office is just right for the work I do. | 3.75 | 3.57 |
| I am generally satisfied with my job here. | 4.38 | 4.28 |

**Table 3.** Significant differences between questionnaire #1 and questionnaire #2.

|  | Questionnaire #1 | Questionnaire #2 |
|---|---|---|
| I would prefer to have my own plants in my office rather than commercially installed plants. | 2.98[1] | 2.63 |
| I would prefer to care for the plants in my office space instead of having them maintained. | 2.63[1] | 2.14 |
| I notice the plants in an office building. | 1.70[2] | 1.86 |

[1]Means were calculated based on a 5-point scale.
Response choices were: 1 Strongly disagree
                         2 Disagree
                         3 No opinion
                         4 Agree
                         5 Strongly agree
[2]Means were calculated based on a 2-point scale.
Response choices were: No
                         Yes

the time of the first questionnaire. Descriptive data show that after plantscaping on the eleventh floor, 14 people out of 30 from the ninth floor said they went to the eleventh floor specifically to see the plants. There were significant differences between males and females, smokers and nonsmokers, people with offices on the ninth floor and those on the eleventh floor, and people who had plants removed and those who did not have plants removed. Positive attitudes towards plants are found throughout our subject population.

Unfortunately, as with any applied field research, complications can arise that challenge methodological rigor. This project was no exception. First, we had communication breakdowns. This project was planned as a cooperative effort between two research teams, yet these two teams were not allowed to contact each other. Second, we failed to identify the most responsible individuals within the organization with the authority to assist in the project. Lack of authority to carry out the necessary procedures obviated systematic collection of verbal report data and hindered the distribution and collection of the first questionnaire. Identifying and working with the administrative services manager before the second questionnaire was distributed resulted in an increased return rate. The site selection was also problematic and not optimal for a study of positive environmental changes. The site for this research was selected for the indoor air quality study and is a healthy, relatively new, attractive building. In addition, it has two styles of floor design. Thus, it was not ideal for a psychological study. A preferable site would have a more uniform office space design on each floor, with a need for some aesthetic improvement. Future research should address the limitations of the measurement instrument and obtain behavioral data to further support the preference studies. Focus groups with employees could also enhance the information available on the issue of plantscapes in the office. They would tap valuable employee opinions about the positive and negative aspects of an office plantscape and facilitate the development of measurement devices.

## LITERATURE CITED

Horsbrough, P. 1972. Human-plant proximities: A psychological imperative. Indiana Nursery News. 33(4):1–4.

Kalmbach, K. L. and J. J. Kielbaso. 1979. Resident attitudes toward selected characteristics of street tree plantings. Journal of Arboriculture 5(6):124–129.

Kaplan, R. 1978. The green experience. In: S. Kaplan and R. Kaplan. (eds.). Humanscape: Environments for people. Wadsworth Publishing, Blemont, California.

Kaplan, R. 1985. Nature at the doorstep. Journal of Architectural Planning and Research 2:115–127

Laviana, J. E. 1985. Assessing the impact of plants in the simulated office environment: A human factors approach. Unpublished doctoral dissertation. Kansas State University, Manhattan, Kansas.

Lewis, C. A. 1972. Public housing gardens—landscapes for the soul. Landscapes for Living, USDA Yearbook of Agriculture, Washington, D.C.

# CHAPTER 17

# Clearing the Air: Horticulture as a Quality-of-Work-Life Intervention

Reginald Shareef

President, SCORE! Management Consultants, Roanoke, Virginia

## INTRODUCTION

Cleaner air is a top priority on the public policy agenda. A new Clean Air Bill is in the last stages of legislative debate. Indoor air quality appears to be the next major environmental issue. Ironically, the toxins we breathe indoors may be more harmful to our health than impure outdoor air. Congressman Joseph Kennedy (1989) has stated that "Americans spend an average of 90% of their time indoors and the air we breathe in schools and workplaces can be more toxic than the outdoor air." Congressman Kennedy and Senator George Mitchell (D., Maine) introduced companion bills in the House and Senate last year to address the problem of contaminated indoor air quality.

Common indoor pollutants include environmental tobacco smoke (ETS), radon gas, and volatile organic compounds (VOCs), including formaldehyde. Pesticides and rodenticides have also been shown to emit toxic "off gases" that are harmful to human health. The medical effects of these contaminants on the American workforce have been dramatic and range from lost productivity due to minor illness to cancer/Legionnaire disease mortalities.

Quality-of-work-life (QWL) applications are organizational attempts that enhance both employee well-being and productivity outcomes. Many QWL approaches target worker environmental concerns that hamper job satisfaction. Clearly, poor air quality can be considered a major workplace environmental issue affecting employee QWL.

Most solutions attempt to remedy this problem by concentrating on improved ventilation systems; however, a new concept in improving air quality has shown that plants are an effective way of removing common pollutants. Citing research conducted at the National Aeronautics and Space Administration (NASA) (Wolverton et al., 1989), the *Knoxville Journal* (Dosser, 1989) reported, "The plants and associated microorganisms will literally consume these toxic chemicals as a food source and convert them into harmless plant tissue."

This paper supports the position of using horticulture as an air purifier in the workplace and presents a methodology for assessing the effect of horticulture on common QWL measures.

## SICK BUILDINGS

Because of the energy crunch of the 1970s, buildings were designed to maximize energy efficiency in an attempt to control escalating fuel costs. Two of the principal redesign features included superinsulation and reduced fresh air exchange (Wolverton, et al., 1989). An unintended consequence of these "sealed" buildings was the build-up of indoor contaminants in office buildings. A World Health Organization Committee recently estimated that approximately 30% of new and remodeled buildings are "sick."

Sick building syndrome (SBS) presents a major threat to worker QWL. SBS illnesses include (but are not limited to) headaches, dizziness, nausea, and fatigue (EPA, 1988).

## THE HONEYWELL REPORT

In 1985, Honeywell, Inc. sponsored a nationwide survey that assessed worker attitudes concerning indoor air quality. The report found that 25% of respondents complained of poor indoor air quality and 20% felt that air quality affected job performance.

These outcomes suggest that a sizeable portion of the sample population was dissatisfied with air quality in the workplace. No questions were asked concerning the medical effects of poor air on employee health. Since nearly 19% of the respondents indicated that they suffered from illnesses that are closely associated with SBS, however, it is safe to assume that workers' QWL and productivity are adversely affected by poor indoor air.

## NASA'S HORTICULTURE RESEARCH

Much of the groundbreaking research utilizing plants in air purification at NASA has been conducted by Dr. Bill Wolverton. This research has been on-going for approximately 15 years and has produced empirical results that show how plants interact with and eliminate toxins in the environment.

Dr. Wolverton and his colleagues studied the effects of plants in reducing the toxicity of three commonly used chemicals: benzene, trichloroethylene (TCE), and formaldehyde. All are potential carcinogens and are hazardous to human systems. The experiments consisted of three steps: (*a*) the injection of the various chemicals (benzene and TCE less than 1 ppm, formaldehyde 6–10 ppm) into a sealed experimental container; (*b*) withdrawal of 200 ml volume of air after a 24 hour interval; and (*c*) immediate collection and analysis of air samples (Wolverton et al., 1989).

Full plant foliage was utilized. English ivy, Janet Craig, peace lily, and marginata were found to be most effective in eliminating TCE and benzene from the sealed environment (Wolverton et al., 1989). Mass cane, pot mum, Gerber daisy, and Warneckei all removed a minimum of 50% of the formaldehyde from the container. This research found that activated carbon filters placed in the sealed chamber also assisted the air purification process.

## PROPOSED METHODOLOGY

I propose a study in which the objectives would be to determine whether plants improve worker QWL in the areas of job satisfaction, intended turnover, and absenteeism. The research methodology would consist of the following three variables:

1. Site selection—any office building declared "sick" by the EPA.
2. Sample—workers in "sick" buildings on different floors.
3. Design—experimental, control-group, time-series.

Workers would be given the Michigan Organization Assessment Questionnaire (MOAQ) before, during, and after plant intervention. This QWL survey instrument has been verified for reliability. Analysis of data should be assessed at six-month intervals via an F-test (ANOVA). One group of employees would have plants in their work environment whereas the other group would not. Additionally, semistructured interviews with a random sample of participants are suggested. The plants would be maintained by qualified horticulturists.

## IMPLICATIONS

Contaminated air quality is extracting a terrible cost on employee QWL and organizational productivity. The cause of this growing problem—sick buildings—does not seem to be lessening, as conserving energy is still a primary concern for most office building owners. Continued field research is now needed to ascertain further the effectiveness of horticulture as an air-cleansing agent in the workplace.

## LITERATURE CITED

Environmental Protection Agency. 1988. Indoor air facts—sick buildings. Report #4. U.S. Government Printing Office, Washington, D.C.

Honeywell, Inc. 1985. Indoor air quality: A national survey of office worker attitudes. Honeywell, Minneapolis.

Kennedy, J. P. 1989. Statement of Congressman Joseph P. Kennedy II upon introduction of the Indoor Air Quality Act of 1989. Office of Congressman Joseph P. Kennedy II, Washington, D.C.

Dosser, R. 1989. Plants found to rid homes of air pollutants. The Knoxville Journal. p. 1. October 24.

Wolverton, B., A. Johnson, and K. Bounds. 1989. Interior landscape plants for indoor air pollution abatement. Report. National Aeronautics and Space Administration, Stennis Space Center, Mississippi.

Rhonda Roland Shearer, *Give Me Shelter,* 1990, Bronze (Lost Wax, Fabrication), 63″ × 43″ × 43″, Emerald Green, Magenta Patina

Rhonda Roland Shearer, *For Sail on a Sunless Sea,* 1990, Bronze (Lost Wax, Fabrication), 85 ¼″ × 75″ × 21″, Red/Orange Patina

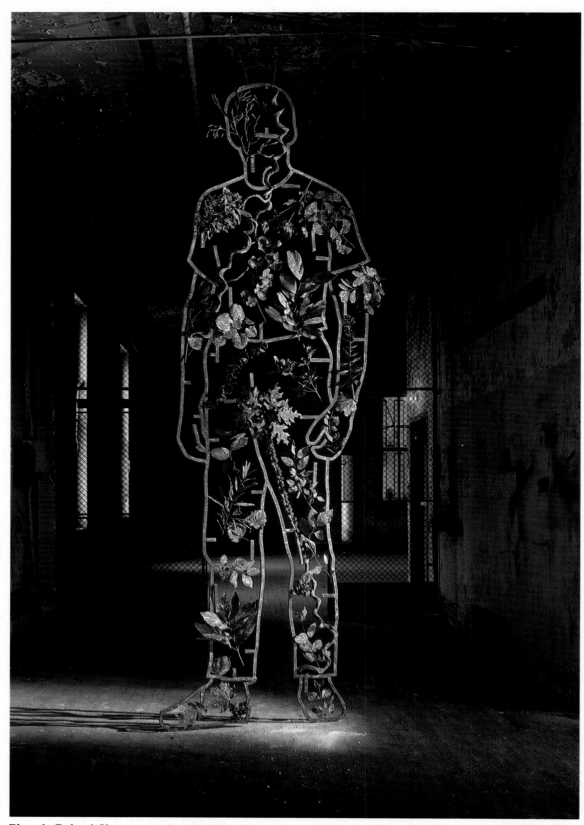

Rhonda Roland Shearer, *Anthropocentrism Series: I Am a Technocrat*, 1990, Bronze (Lost Wax, Fabrication), 144″ × 43″ × 5½″. Red Patina

Rhonda Roland Shearer, *Birth Chair*, 1990, Bronze (Lost Wax, Fabrication), 41¼″ × 25″ × 21½″,
Yellow/Green Patina

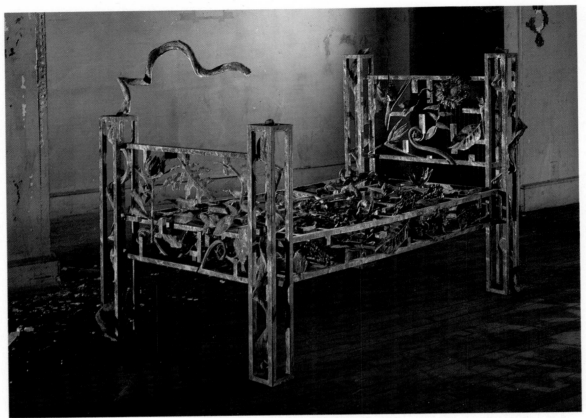

Rhonda Roland Shearer, *Sleepless*, 1990, Bronze (Lost Wax, Fabrication), 44″ × 66″ × 37¼″, Purple Patina

Rhonda Roland Shearer, *The 5 Platonic Solids*, 1990, Bronze (Lost Wax, Fabrication): *Terra*, 34¾″ × 33½″ × 34¾″, Mirror, Blue Patina; *Aqua*, 46¾″ × 55″ × 55″, Red Patina; *Ignis*, 38½″ × 39½″ × 39½″, Yellow Ochre Patina; *Caelum*, 46″ × 55″ × 54½″, Viridian Green Patina; *Aer*, 37″ × 40″ × 40″, Orange Patina

CHAPTER 18

# A Quantitative Approach
# to the Description of the Qualities
# of Ornamental Plants,
# with Particular Reference to Plant Use
# in the Rural Environment

---

Marion B. MacKay

Lecturer in Amenity Horticulture, Massey University, New Zealand

David J. Chalmers

Professor of Horticultural Science, Massey University, New Zealand

## INTRODUCTION

The ability to facilitate positive change in the environment requires knowledge and understanding of the interaction among people, plants, and the environment. Knowledge of the individual elements of the relationship is required to understand their interaction. This paper presents a progress report on a research program that considers some of these elements and interactions.

Two important levels of interaction between user and environment may be identified. First, the user interacts subconsciously with the landscape when using or moving within its masses and spaces. At the second level, the user relates directly to plants in the landscape; this involves a process of recognition and appreciation leading to some action relating to that plant. In this research, we are interested in the interaction between the conscious user and individual plants in the context of the nature of the New Zealand landscape and the extent of the ornamental tree resource in New Zealand.

Agricultural practices have dominated the rural New Zealand landscape. Intensive land clearing to create pasture formed a landscape in which trees were only considered for utilitarian purposes. Where trees were used, the choice of plant material was directed towards wind-hardy material that grew rapidly, consequently a narrow range of exotic species, including *Pinus radiata*, *Cupressus macrocarpa*, *Populus*, and *Salix* species, were repeatedly used and now dominate the scene. Much of the landscape therefore lacks variety in trees, especially the long-lived and slower-maturing species.

Agricultural practices have also led to severe land degradation problems in some areas. Tree planting has been used to some extent to control this problem, stimulated to a degree by large subsidies and incentives. Again, the functional qualities of the plants used for this purpose dominated the choice of species, thus leading to further planting of *Populus* and *Salix* species.

The nature of the New Zealand landscape thus described reflects attitudes and preferences of the community regarding the use of trees in the landscape and the appearance of the landscape as a whole. Individuals contribute to the general landscape by the selection and use of plants in the part of the landscape for which they are responsible.

New Zealand has a mild climate that is favorable for the growth of many plants, and the range of tree species in New Zealand is diverse. This diversity, however, is under utilized. Many species are held in specialist collections that are not available to the public user. At the same time, contemporary marketing pressures in the ornamental plant industry have led to a focus on colorful, often trivial, plants that are usually transitory elements. There is a danger that these marketing approaches may exacerbate the loss of larger and longer term elements over time, thereby degrading the landscape.

These factors underpin the philosophy behind our research program. We propose to consider a range of users and their interactions with a diversity of plants. Plants will be assessed for their landscape values by groups of people with varying expertise and their reaction and preference to specific plants quantified. We propose that the knowledge gained from this study can be developed into a capacity for positive influence on the landscape and may ultimately be used to raise awareness of the use of trees in the landscape, thereby leading to significant environmental improvement.

## RESEARCH PROPOSAL

A model is being developed to study and utilize the interaction between people and the plants they use. The perception and preference of a group of expert horticulturists has been quantified to develop a management philosophy for a horticultural resource. In the future, similar models will be used to evaluate the needs of nonexpert groups using the same resource. The model is based on the assumption that an objective method of assessing the qualities of ornamental plants can be formulated that analyzes an individual's perception of and reaction to plants. The needs of a group, thus defined, can then become part of the decision-making process to manage plants in the landscape.

## MODEL DEVELOPMENT

The relationship between plants and people exists in a continuum of users from the totally uneducated to the expert user. One might, therefore, identify botanists, interested gardeners, children, or the elderly as discrete assessable populations. Different groups could be expected to be associated with different issues in relation to the landscape and its utilization. This model assumes that by grouping users of the environment into broad categories, the needs of that group can be identified and analyzed. In this project we focused on a group of expert horticulturists. Accordingly, their perception of plants was sought in terms of botanical and aesthetic value of plants on a test site. Our aim was to assess and quantify the opinion of the expert group, and thereby to develop a management policy for the composition, maintenance, and development of the test site.

The test site was the Eastwoodhill Arboretum in New Zealand, which is that country's most comprehensive collection of ornamental trees and shrubs. The diversity and maturity of the collection make it ideal for this type of study. Many plants in the collection are no longer in common use, and therefore the site affords a significant opportunity for the examination of a wide range of examples, while the maturity of trees allows them to be considered in their ultimate form.

The need for planning and development at the test site was a key factor in the development of the model. Both scientific and public users of the resource at a number of levels had to be accommodated. The arboretum has been maintained in an ad-hoc manner for all but the last few years, thus creating an immediate need for botanical and scientific planning.

## THE MODEL

In the first step (see Table 1), participants gave a measure of their opinion of the botanical and aesthetic merit of each plant by allotting a score on a defined scale. This drew on their expert knowledge of the plant species in question but did not refer to any particular specimen and was completed away from the arboretum. The scores generated from the first assessment allowed initial prioritizing of plants in terms of botanical and aesthetic importance.

Once on site, opinion was sought from participants on the characteristics of the example of the specimen growing in the arboretum. Plants were again rated but this time for health, individual aesthetic appeal, and contribution to the landscape scene. The scores given from this assessment allowed plants to be evaluated for their suitability as representatives of the species at the arboretum.

In step three, the two sets of data were combined. Discussion resulted in a decision for each plant and a management proposal for the study area.

The results given in this paper report the development of the model and the use of the model to derive a successful solution to a management problem at the arboretum. A solution was facilitated that was based on the objective assessment of the opinion of the expert group in relation to the qualities of the plants in question (see Table 2). This approach has now been applied to two problem sites within the arboretum.

**Table 1.** A model for using expert opinion on plants for decision-making in landscape design

| Stages of prioritizing | Selected test plants | Solicited opinion |
| --- | --- | --- |
| First prioritizing | Trees of importance highlighted | Off site Botanical and aesthetic merit |
| Second prioritizing | Individual examples highlighted | On site visit Health rating; landscape contribution |
| Third prioritizing | Decision on each plant | Discussion Availability of each species |
| Results | Management proposal for the area | |

**Table 2.** An example of tree management decisions based on assessment of trees by using expert opinion.

| Tree | Step one | | Step two | | Step three (Decision)[a] |
|------|----------|---|----------|---|--------------------------|
| | Botanical | Aesthetic | Health | Landspace contribution | |
| *Abies amabilis* | 8.0 | 8.0 | 2.8 | 3.1 | Remove |
| *Abies bracteata* | 8.0 | 7.5 | 9.7 | 10.0 | Retain |
| *Abies* 'Candicans' | 9.0 | 10.0 | 2.6 | 4.2 | Propagate, remove |
| *Abies georgiana* | 8.0 | 6.0 | 4.0 | 4.8 | Remove |
| *Abies holophylla* | 9.0 | 7.0 | 9.1 | 8.4 | Retain |
| *Abies veitchii* | 8.0 | 9.0 | 8.2 | 8.8 | Retain |
| *Picea bicolor* | 9.0 | 6.0 | 6.0 | 6.2 | Retain |
| *Picea morrisonicola* | 9.0 | 7.5 | 8.4 | 8.6 | Retain |
| *Picea obovata* | 9.5 | 6.5 | 1.5 | 4.0 | Retain, propagate |
| *Picea orientalis* | 7.5 | 8.0 | 9.5 | 9.5 | Retain |
| *Picea spinulosa* | 8.0 | 9.0 | 6.2 | 7.1 | Retain |
| *Picea wilsoni* | 8.0 | 6.5 | 3.3 | 4.0 | Remove |

[a]The time available for this paper does not allow a detailed description of how the decision for each plant was reached.

## DISCUSSION AND CONCLUSION

In this simple form, the model allowed the quantitative use of expert opinion in a decision-making process. The approach has subsequently been modified to use the same basic inputs (expert opinion, botanical information, a series of prioritizing steps) but also to allow for changes over time. Thus modified, the approach addresses not only the immediate problems as they arise, but also can predict patterns in park development and therefore facilitates planning for the park objective of high visual appeal while maintaining botanical interest to satisfy a range of users.

In both forms, the model brings together user opinion and plant qualities and synthesizes these to give an objective basis to landscape decision-making with respect to the use of plants. At the same time, the users are making a contribution to the development of the landscape and are therefore increasing their perceived value of the landscape resource.

CHAPTER 19

# The Contribution of Interior Plants to Relative Humidity in an Office

Virginia I. Lohr

Assistant Professor of Horticulture and Landscape Architecture,
Washington State University

The relative humidity of the air inside buildings can be extremely low. This is especially true when buildings are being heated, because the relative humidity drops as air is heated if no supplemental moisture is added. Relative humidity is the amount of moisture in air expressed as a percentage of the maximum amount the air is capable of holding, and warm air can hold more water than cooler air. The relative humidity in heated buildings is often below the recommended 30–60% range for human comfort (Price, 1984). In this paper I review the positive and negative effects of relative humidity and discuss the effects of adding plants to humidify air.

High relative humidity in buildings causes numerous problems, such as moisture condensing on cold surfaces, including windows and exterior walls in winter, and structural damage can result (ASHRAE, 1989). Many molds and mildews are prolific when the relative humidity exceeds 75% (Anderson and Korsgaard, 1984). These can damage surfaces and cause health problems. Allergic diseases like asthma and rhinitis can be triggered by exposure to proteins from house dust mites, which grow faster in homes with higher relative humidities (Anderson and Korsgaard, 1984).

The relative humidity inside buildings should be maintained below the point of saturation to prevent damage from high moisture. Buildings are routinely designed to remove humidity by venting interior air to the outside, through processes as complex as air exchangers or as simple as leakage through walls and openings (ASHRAE, 1989). If interior and exterior air were not routinely exchanged, the interior relative humidity would rise to saturation, because there are many sources of moisture in most buildings: for example, people release moisture through their skin and as they breathe, and cooking and washing add moisture to the air.

When the relative humidity of interior air is too low, other problems can be created.

Wood dries out, warps, and cracks at low relative humidity, and extreme variation in relative humidity creates problems as wood expands and contracts, causing cracking and creaking (Hoadley, 1980). In studies conducted at relative humidities between 15 and 55%, there was evidence that human colds were more frequent at lower relative humidities (Green, 1984). Highly controlled studies of the effects of humidity on the incidence of human colds are difficult for a number of reasons; for instance, most people spend time in several locations during a day, and each location may have a different humidity level. Studies on the incidence of influenza virus infection in mice revealed a pattern of increased infection at lower relative humidity (Kilbourne and Schulman, 1962, reported in Green, 1984).

The routine venting of interior air to reduce humidity can cause the relative humidity level to drop excessively during cold weather if no systems are in place to supplement moisture. This occurs because as exterior air is heated, the amount of moisture it can hold increases, so the relative humidity drops. Most buildings do not have systems to maintain humidity within desirable ranges. In those that do have systems to humidify the air, humidifiers can become contaminated with microorganisms that cause human disease (Green, 1984).

Plants contribute to increased interior humidity primarily by adding moisture to the air through transpiration through their stomates and secondarily through evaporation from growing media and drainage dish surfaces. Those who design buildings are concerned that plants may contribute excessive amounts of moisture to interior spaces (ASHRAE, 1989). Although the amount of moisture added by plants cannot be precisely controlled the way that moisture added by humidifiers can, natural variability in the rate of transpiration leads to some self-control. When the relative humidity is high, the transpirational driving force (vapor pressure deficit) is reduced, so the rate of transpiration also is reduced over what it would be at lower relative humidities.

The amount of moisture that plants contribute to surrounding air is influenced by many factors, including plant species, the moisture content of the growing medium, temperature, light levels, and previous growing conditions (Kramer, 1983). The amount of moisture that plants contribute can be estimated from the amount of water applied to the growing medium. Whether this will have a significant effect on relative humidity will depend on the initial relative humidity, the volume of air in the space enclosing the plant, and on the rate of air exchange for that space. Generally, newer residences with reduced air leakage rates and older commercial buildings without central, forced-air systems will see the greatest increase in humidity from adding plants, because the rate of air exchange in these spaces is low. In commercial buildings, the humidity contribution from plants in one space will tend to be distributed throughout the building, thus reducing the impact in the space where the plants are located.

To determine the potential impact of plants on relative humidity in a single office space in a building with a central, forced-air system, I recorded the relative humidity in the presence and absence of plants. Measurements were conducted between December 1989 and March 1990. Once each week, the treatment condition was randomly assigned and plants were added or removed as required. Humidity and temperature were recorded every six hours. The office space was about 35 cubic meters. Plants filled about 1.25 cubic meters of the interior space and used between six and eight liters of water each week. A variety of plant species (ferns, palms, peperomia, and aglaonema) were used. Air exchange rates from the space were not measured directly but were estimated to average one to two air changes per hour.

During the experiment, the average office temperature was 22°C and did not vary by treatment. Average relative humidity in the absence of plants was 25% and in the presence of plants was 30%; these values were significantly different at greater than a 0.01 level. The magnitude of the increase, as expected under these experimental conditions, was not large. It was probably due to air exchange within the building, and higher humidity would be attained by adding more plants to the single office or by adding plants to other offices within the building.

More careful selection of the species, size, and condition of the plants might result in a larger increase in relative humidity. For this experiment, most plants were purchased and repotted immediately prior to the experiment; they may have been under stress from reduction in light levels from production conditions to office conditions or water stress from root disturbances during transplanting. Another possible confounding factor was the presence of carpeting and upholstered chairs, which can absorb moisture when relative humidity is high and release that moisture slowly over a period of weeks (ASHRAE, 1989).

These results showed that plants added to an office in a building with a forced-air system, which redistributes air within the building, could significantly increase the relative humidity in that office. Relative humidity in the absence of plants was slightly below the range recommended for human comfort, whereas it was within the recommended range in the presence of plants.

## ACKNOWLEDGMENT

H/LA Paper No. 90-09. Project No. 0695, College of Agriculture and Home Economics Research Center, Pullman, WA 99164.

## LITERATURE CITED

Anderson, I. and J. Korsgaard. 1984. Asthma and the indoor environment: Assessment of the health implications of high indoor air humidity. Proceedings of the Third International Conference on Indoor Air Quality and Climate. Volume 1: Recent advances in the health sciences and technology, pp. 79–86.

ASHRAE. 1989. ASHRAE handbook fundamentals. American Society of Heating, Refrigerating and Air Conditioning Engineers, Atlanta, Georgia.

Green, G. H. 1984. The health implications of the level of indoor air humidity. Proceedings of the Third International Conference on Indoor Air Quality and Climate. Volume 1: Recent advances in the health sciences and technology, pp. 71–78.

Hoadley, R. B. 1980. Understanding wood. The Taunton Press, Newtown, Connecticut.

Kilbourne, E. D. and J. L. Schulman. 1962. Airborne transmission of influenza virus infection in mice. Nature, Sept. 15: 1129. Taken from: Proceedings of the Third International Conference on Indoor Air Quality and Climate. Volume 1: Recent advances in the health sciences and technology, p. 73.

Kramer, P. J. 1983. Water relations of plants. Academic Press, New York.

Price, S. 1984. Indoor air quality. Washington State Energy Office, Washington Energy Extension Service Bulletin EY3200.

# The Relationship of Plants to Lifestyle and Social Support

Kim Randall,* James E. Healy,* Diane Relf, † E. Scott Geller ‡

*Graduate Student, Department of Psychology, † Associate Professor of Horticulture, ‡ Professor of Psychology, Virginia Polytechnic Institute and State University

## ABSTRACT

The Lifestyle and Social Support Questionnaire, a test designed to measure various sources of social support and risk behaviors for university screening, incorporated four questions about people's interaction with and opinions about plants. This questionnaire was then administered to 483 students at a large university during the summer session. The results of this study indicated that most people like plants.

# The Impact of Floral Products in Restaurants

Kevin L. Grueber

Assistant Professor of Horticulture, Virginia Polytechnic Institute and State University

Michael R. Evans

Associate Professor of Hotel, Restaurant, and Institutional Management,
Virginia Polytechnic Institute and State University

## ABSTRACT

Many restaurants provide floral or plant products on or near tables in restaurants, presumably to improve the ambiance. Such products may be live, silk, or plastic, and may vary in quality, size, and their impact on the customer. We describe a research project that explores the effects of floral products on customer attitude and behavior. Potential products include live flowers of variable quantity and quality, live plants, and artificial flowers. Potential impacts include customer satisfaction, purchasing, tipping, and repeat business. We also discuss methods of assessing economic gain/loss and of returning the gathered information to the restaurant and floral industries.

# SECTION IV

# *DEVELOPING A CONCEPTUAL FRAMEWORK*

CHAPTER 21

# The Psychological Benefits of Nearby Nature

---

Rachel Kaplan

Professor of Environmental Psychology, University of Michigan

## INTRODUCTION

That plants are important to people is evident in many ways. Flowers are often a center piece of joyous occasions. They are no less dominant at many sad events. The significance of plants to people's well-being knows no demographic boundaries; the relationship holds for different cultures, ethnic groups, levels of affluence, occupations, ages, amount of education, place of residence, country of birth. This very volume and the national symposium it represents are further testimony to this significant relationship.

To cast this relationship in terms of "plants," however, may diminish our understanding. The focus of this paper is on a somewhat broader concept—nearby nature. An explanation of what is subsumed by this concept and the rationale for focusing on it (as opposed to "plants" or "horticulture") are the objectives of the first part of the paper. The middle section concerns the ways people relate to nearby nature—their uses of it, the satisfactions they derive from it. Neither plants nor people are all alike, however. The final section thus considers some differences among people and the implications that such differences suggest for enhancing well-being.

Although these three themes are relatively straightforward, it is impossible to do them justice in these pages. The anecdotal evidence and indirect indications of the importance of nearby nature are abundant. Many of us, however, continue to be surprised by the dearth of research in this area. Fortunately, in the last few years this important subject has received more empirical attention. Several of the chapters in this volume report on such work.

To reduce duplication, I will rely most heavily on our own research here. Even so, the present chapter necessarily glosses over many important questions. *The Experience of Nature* (Kaplan and Kaplan, 1989) provides a fuller discussion of some of these as well as references to many pertinent studies.[1]

## THE EXPERIENCE OF NATURE

"Nearby nature" includes one plant or many plants, and also the place created by them. It includes a street tree as well as the trees in an atrium. A parsley plant on the window sill and an herb garden are both also part of "nearby nature." So are an arboretum, a person's well-tended garden, or a less nurtured "backyard." We also include in this concept nearby fields and woods and land that has not yet been turned to development.

In other words, nearby nature is about vegetation that is proximal. It can be indoors or out-of-doors; often it is outside but viewed from inside. It can be flowers and plants as well as settings that include plants. A rose bud in a vase and an arboretum certainly range widely in scale. A street tree and a neighborhood park range widely in pertinent activities. One's own garden and someone else's flower box differ substantially in the involvement of a particular individual.

Such a broad definition runs contrary to the accepted terminology of many professional groups. For some, "nature" is reserved for wilder places and does not exist in the urban context. For others, there may be "nature" in the city, but it is what bigger parks are about. "Nature" and "horticulture" also express very different domains to professionals involved in recreation, landscape architecture, and botany.

Flora or vegetation or natural settings are the province of many distinct professional groups. For them it is reasonable and important to maintain such different perspectives. The responsibility of maintaining an indoor palm tree or bamboo grove in an urban atrium demands an expertise different from the requirements of designing a vest pocket park. The focus on "plants," or "flowers," or "parks" as separable domains is thus understandable.

The reason for emphasizing the broader concept of "nearby nature" thus requires justification. We must examine both whether there is a rationale for combining such diverse domains and whether there are gains from doing so. The key to both these issues lies in how the natural environment is experienced by people in their ordinary lives (as opposed to their professional roles) and the benefits they derive from their experience.

### The Perception Puzzle

Professional training involves learning to see things in a particular way. A plant pathologist and a bonsai specialist are likely to see the same plant in distinctly different ways. Prior to their training, however, their experience of the same plant would have been different from either of their current ways of seeing. The acquired way of seeing is essential to the process of becoming an expert. In gaining such expertise, one not only learns a vocabulary that is shared by that professional group, but also a way of making distinctions, of recognizing salient characteristics, of understanding.

Experts are not usually aware that they see things differently. The processes of recognizing and categorizing generally occur effortlessly and all of us assume that our perception is no different from anyone else's. This is equally true in areas where our "training" is not formal, but rather the result of extensive daily experience. Thus people learn to recognize and categorize different kinds of settings based on repeated encounters. Doing this is a routine part of perception and is carried out without our awareness much of the time.

Since people are not aware of their perceptions, it is difficult to ask about them directly. To study how people experience the natural environment, however, it is necessary to know about these invisible perceptions. Our research program for the last two decades has addressed this issue through an intermediate concept: *preference*. People have no difficulty whatever in indicating how much they like something. By analyzing the patterns of these judgments, it is possible to learn about perception or categorization. Thus the preference judgments are not only useful in their own right, as an indication of the kinds of things or settings that people favor, but also as a way to understand how the things or settings are experienced.

## The Nature Category

People react to what they experience in terms of commonalities, in terms of classes or categories. A scene is generally perceived as a particular instance of a larger class of scenes. By asking people to indicate their preference for each of many scenes (which have been carefully selected to sample particular kinds of settings), one can determine some of these commonalities. By then comparing the results of numerous studies, it is possible to begin to understand how different kinds of settings are experienced.

The basic approach we have taken, then, involves using photographs or slides of different kinds of settings and asking study participants to rate each scene on a 5-point preference scale. This is a task that people of different cultures, ages, educational levels, and backgrounds have all performed easily and enthusiastically. Statistical procedures are available for extracting the common themes, based on the pattern of responses. We have called these procedures *category-identifying methodology* (CIM) because they indicate which scenes "belong" to a common theme or category. The interpretation of what the underlying common elements are for each category, however, cannot be achieved by mechanical computation. These are the responsibility of the researcher.

*The Experience of Nature* presents synopses of dozens of studies that have used this approach. The findings of each of these studies address some specific kind of setting, such as roadsides, common urban settings, or residential environments. An analysis of the kinds of categories that have emerged across the various studies is the basis for the idea of "nearby nature." In other words, it is from such analyses that we have come to the conclusion that "nature" is a critical component in how people experience the environment.

The categories that emerge from the various studies fall into two major types. One of these can be described as based on *content;* the other focuses on *spatial configuration.*

A major underlying theme in many of the content-based categories concerns the balance between human influence and the natural area. Thus the perception of settings is strongly influenced by the degree and kind of human intrusion. Scenes that are dominated by buildings tend to form discrete categories, as do scenes that are relatively low in such influences. The presence of a road, or cars, or telephone poles does not detract from the perception that the scene is largely natural. (These items are, however, likely to lead to lower preference judgments.)

Even more striking than the particular content domains, however, are the categories for which content is not the distinguishing characteristic. In these cases, it is the spatial configuration or organization of the scene that appears to account for the categorization. The "space" in question here is not the two-dimensional space of the picture plane, but the inferred three-dimensional space of the scene that is depicted. These categories suggest that an underlying criterion in making a preference judgment is an evaluation of the scene in terms of the presumed possibilities for action, as well as the potential limitations.

The spatial configurations categories can be further distinguished in terms of the degree of openness and the extent of spatial definition. The "wide-open" theme has been evident in the CIM results of many studies. The scenes comprising this category generally lack any particular differentiating characteristics and the sky occupies a considerable portion of the scene. Farmland, unused roadsides, bogs, marshes, and shorelines all provide examples. Scenes that *lack* openness also emerge as a separate category. Here too there is a lack of focus or of differentiating characteristics, but rather than giving a sense of endless open space, the view is blocked.

The categories that are strong on spatial definition can often be characterized as "parkland." These are settings that are relatively open, but have some distinct trees that greatly enhance the sense of depth. As a group, these categories tend to be among the most highly preferred kinds of natural settings.

## Summary

The results of our research suggest that whereas plants are of great importance to people, the specific plants are not the major focus of the way people experience the environment. Essential to perception is the presence of vegetation and the context created by it. The degree to which the setting is natural and the arrangement among the elements of the setting are particularly salient in people's implicit categorization (R. Kaplan, 1991).

Thus the broad, somewhat nebulous designation of "nearby nature" reflects the breadth of the human experience. Of course people can, and frequently do, make distinctions among the various kinds of settings that are included in this concept. A flower arrangement is not a park. The point, however, is that the kinds of experiences people have and the benefits they accrue from the different kinds of "nearby nature" have some striking similarities. Examining the broader concept, therefore, may be more useful for understanding the nature of the human experience.

## THE NATURE OF THE EXPERIENCE

Many indirect sources of evidence point to the role that natural settings play in satisfaction and well-being. One can, for example, use dollar expenditures as an indicator. People spend great sums of money for plants, flowers, landscaping, and recreational pursuits. One can also use time expenditures to gauge importance. Many hours are devoted to hobbies and activities that are nature-related. Other indirect measures can be derived by examining the kinds of settings that are attractive to tourists and by looking at how people arrange their home environment.

It is useful to consider the kinds of involvement that people seek with respect to plants and natural settings. One would assume that active recreational pursuits, in a nature context, would be beneficial. After all, people seek opportunities for being in natural places. Hiking and gardening are among the most popular outdoor pursuits, in terms of the numbers of people who participate.

Perhaps less obvious is that many of the benefits from nearby nature do not entail such active involvement. Examination of the research that has demonstrated the important role that the natural setting plays in satisfaction and well-being points to two forms of involvement that seem particularly salient. One of these involves opportunities for noticing or observing; the other derives from the knowledge that the opportunity is available—even if it is not "used" or directly seen.

### The View from the Window

A great deal of satisfaction derived from nature does not involve being in the natural setting, but rather having a view of it. It is hardly surprising that studies on windowless settings (including schools, hospitals, work environments) suggest that they are unpreferred (Verderber, 1986). The several studies that have shown health benefits related to nature have all been based on opportunities for noticing and observing, rather than on activities that are performed in nature. Moore (1981) and West (1986) both reported that prison inmates used health-care facilities significantly less often if the view from their cells was toward natural areas. Ulrich (1984) and Verderber (1986) found recovery in the hospital setting to be significantly enhanced by the quality of the view from the patient's room.

Residential satisfaction has also repeatedly been shown to be related to the availability of nearby nature. Fried's (1982, 1984) work is particularly noteworthy in this context. In a carefully drawn, national survey, he found that the strongest predictor of local residential satisfaction was the ease of access to nature. This was also the second most important factor (after marital role) in life satisfaction. The pattern of these findings was particularly strong at the lowest status levels. Thus, for lower social class positions, the satisfaction with the physical setting is even more powerful in explaining life satisfaction than is the case as social status increases.

Similar results, though not from a national survey, were reported by Frey (1981), who found neighborhood satisfaction strongly affected by the availability of nearby nature. Neighborhood satisfaction, in turn, was a significant aspect of participants' perceived degree of life satisfaction.

Neither Fried's nor Frey's studies specifically address the view from the window. It is to be expected, however, that much of the satisfaction derived from the nearby trees and natural areas stems from seeing them from one's dwelling. A study based on nine multiple-family housing projects (Kaplan, 1983) asked specifically about the view from home. Participants' satisfaction with their community was strongly related to having views of gardens; views of woods and trees were particularly important factors in several other neighborhood satisfaction measures.

The importance of opportunities to see nature is not limited to special populations (such as long-term hospital patients or inmates), nor to the home setting. The role of nearby nature in the work environment has received only minimal empirical attention. Our initial study in this context (S. Kaplan, et al., 1988) found that workers with a view of natural elements, such as trees and flowers, experienced less job pressure and were more satisfied with their jobs than others who had no outside view or who could see only built elements from their window. Employees with nature views also reported fewer ailments and headaches.

A study we are currently conducting includes 616 individuals whose jobs are largely sedentary. Their positions vary widely in pay and responsibility but have in common that their working day is spent indoors, generally in a limited area. Once again, the results show that opportunities to see natural elements is a strong positive factor in enthusiasm about work and in satisfaction with the work situation.

## Thereness

Consider the following finding in the study about nature in the workplace. The participants were asked how difficult is it for them to see outside from their desk or workstation and how likely are they to look out a window in the course of the day. It is not surprising that both the ease of seeing out and the likelihood of looking out are key issues in the employees' satisfaction with the view from their workspace. Of course, satisfaction is also affected by what they can see, and the ability to see more natural elements enhances the satisfaction. What may be more surprising, however, is the relative importance of these factors. The difficulty of seeing out plays a far more substantial role than does the likelihood of looking out.[2] In other words, knowledge that a view is available is in itself important, even if one does not take advantage of the opportunity to do so.

We have referred to this phenomenon as "thereness," an appreciation of the natural setting by virtue of its availability, whether or not one partakes of it (R. Kaplan, 1978). It is an important aspect of the contribution of nearby nature that is easily overlooked if the evidence one seeks is from aggregate measures such as dollar expenditures or is based on respondents whose active involvement with a natural setting is a condition for their participation in a study.

Thereness is a particularly vulnerable source of satisfaction. When people's appreciation of a resource is based on their cognitive state—knowing that it is there—rather than on actual use, it is all too easy to assume that the resource does not "really" matter. Neighborhood parks often lie vacant, yet people want to live near them. Human beings are readily dismayed by the lack of choices and prosper from hope and opportunity.

## Summary

The psychological benefits that nearby nature offers are based on many forms of involvement. Although being in nature and nature-based activities are important sources of satisfaction, the experience of nature is often derived from far more subtle pursuits. The availability of flowers, plants, trees, and nature places and the perceived adequacy of opportunities to be in contact with nature have been shown in several studies to be important components of well-being. The knowledge that such settings and opportunities to see nature are available to be enjoyed is in itself a source of satisfaction.

## THEME AND VARIATION

The theme of this paper is the importance of nearby nature to human satisfaction and well-being. Anecdotal and more systematic evidence suggests that nature does indeed play a significant role in well-being. Although the pervasiveness and consistency of this relationship is remarkable, it certainly does not suggest a universal pattern without variation. All plants and natural settings are not equivalent in their effects, nor are all people equivalent in their responses to nature. It is important both to recognize the pervasiveness of the importance of nature and to acknowledge the variations on the theme.

The variation in response to nature is closely related to familiarity and experience. This would suggest that the role that nearby nature plays differs in the course of the life cycle. Where one lives, one's cultural heritage, as well as one's travels would lead to different experiences. Furthermore, formal training and expertise can have direct bearing on one's familiarity with plants and natural settings. Although these factors would be expected to make a difference, the research literature that speaks to these issues is scant. Clearly this is an area in need of further investigation.

### Life Cycle

Extensive empirical work is not required to tell us that the very young and the very old differ in their recreational patterns. Activities that involve the nearby natural environment may be of great importance at both ends of the life cycle, and to every age in between, but the likely "uses" or forms of involvement are different.

The experiences highlighted in the previous section—noticing nature and appreciating its "thereness"—are hardly the domain of toddlers. They are evident, however, at many later stages of the life cycle and are strongly expressed among the elderly (Talbot and Kaplan, 1991).

The Balling and Falk (1980) study on preferences for different biomes shows some fascinating age-related variation. The two youngest age groups in their study (ages 8 and 11) showed stronger preference for savanna scenes than for the deciduous and coniferous forests that were more characteristic of their personal experiences. For all other age groups, these three biomes were equivalent in preference. The preferences of the 15-year old group were also noteworthy. They were consistently lower in preference for each of the five biomes included in the study. Medina's (1983) study also showed distinctly lower preferences for nature scenes among people of this age group. The findings are suggestive of a developmental pattern among urban adolescents; they call for further research to answer many questions.

## Background Variables

Traditionally, social scientists include a variety of questions about the study participants' background. These questions establish the demographic composition of the sample and explore whether background accounts for differences in participants' responses to other items. Thus, research on the role of plants in well-being might include questions about the respondents' rural or urban background, length of residence at the current location, ethnicity, income, etc.

From such questions, Fried (1982, 1984) derived the findings about the particular significance of nature access for lower income groups. Similarly, several studies have explored ethnic differences. The results suggest that the theme of the importance of nearby nature holds true across groups, and that variations on this theme exist as well. In particular, the black individuals sampled seem to have a stronger preference for the more managed or manicured, neater and more orderly settings of nearby nature (Kaplan and Talbot, 1988).

Cross-cultural studies have frequently demonstrated strong similarity in environmental preference. This may not be surprising when the cultural and environmental patterns are similar. Yang and Kaplan (1990) found strong similarities even for dissimilar cultures and for distinctly different landscape styles. Koreans and western travelers in Korea were asked to indicate their preferences for scenes showing Korean, Japanese, and Western styles. The category-identifying methodology (CIM), described above, revealed remarkably similar perceptions for the two samples: the Japanese landscapes were the most preferred. Particularly noteworthy was the low preference expressed by the Korean sample for their own characteristic landscape style.

## The Trained Eye

If indeed, as suggested above, professional training involves learning to see things in a particular way, one would expect that those with plant- and nature-related expertise would have reactions different from others'. Extensive experience with horticulture, botany, silviculture, landscape architecture, gardening, turf management, or any of the many other professions and avocations pertinent to this theme would surely not diminish one's sense of the importance of the natural world. Such experience can, however, affect how one sees nature, what one considers important, and the satisfactions one derives. In fact, the role of knowledge or expertise is likely to be a particularly significant source of variation for our central theme. The consequences of failing to acknowledge that expertise makes a difference can have unfortunate, though unintended, consequences.

In *The Experience of Nature* (Kaplan and Kaplan, 1989; chapter 3), we discuss several studies that deal with the role of knowledge in people's responses to nature. Members of special interest groups are likely to differ from the population as a whole in their concerns and priorities. Environmental planners and resource managers differ from citizens in the different categories they use for identifying what needs attention in the natural settings. Managers of residential settings may make decisions that run counter to the preferences of the residents. It is certainly not difficult to think of other examples in which someone else's judgment ran counter to our own, in which someone "in charge" was insensitive to the implications of what was to them a reasonable decision.

It is more difficult, I suspect, to recognize that when we are the experts the same situation occurs. In other words, we are less likely to be aware of the consequences of our own expertise than of others'. We are generally unaware of our perceptions and of the bases for many of the actions we take. Thus, all too easily, decisions are made that undermine the availability of the natural environment, or worse yet, that destroy nearby nature because of a different set of priorities.

Experts are necessary and their contributions are vital. Nonetheless, the limitations inherent in expertise can have harmful (though unintended) consequences. The implication here is not to do without experts, or to do without the contribution of those on whose behalf the expertise is sought, but that participation by untrained individuals who are impacted by the experts' decisions is an essential part of the process.

## Summary

The pervasiveness and consistency of the importance of nature is perhaps more striking than are the variations on the theme. Nonetheless, it is important to acknowledge variations: in the expression of what is important, in preferences for different plants and nature settings, in priorities and urgencies with respect to nature. Rather than assume that all people are alike, or that one group knows what another group prefers or needs, the expression of both the theme and the variation is vital.

## SOME IMPLICATIONS

Nature is many things. It comes in many colors, forms, sizes, and availability. It calls for different forms of involvement, permits different degrees of intensity. It bestows a great variety of pleasure and joy. Why or how nature is beneficial may not be self-evident; that nearby nature is important, however, is unquestionable for many people. It is important in different ways for different people and for the same person on different occasions and in different phases of life.

This is nature nearby, not the distant wilderness, wild rivers, and scenic mountains. This is nature that is unspectacular and ungrand. Unless, of course, one sees grandeur in the changes that the seasons bring, in the opening of a bud, in the way one's garden takes shape. Without a tree nearby, one cannot witness the seasonal variation, one cannot struggle to catch a glimpse of the birds, or watch the antics of the squirrels chasing each other. Without a spot of ground nearby, one cannot help it take shape nor imagine the wonders it will yield next summer.

There are many places where nature is removed to make way for housing. There are many places where people live that offer no view of nature. There are many places where people work that have no plants or even pictures of plants. Such conditions represent, at best, a serious misunderstanding of the role of nearby nature in human well-being. Nature is not merely an amenity, luxury, frill, or decoration. The availability of nearby nature meets an essential human need; fortunately, it is a need that is relatively easy to meet. A garden patch, some trees nearby, and a chance to see them can all be provided at minimal cost and for enormous benefits.

## NOTES

[1]Reference to "our" research program refers to collaboration by Stephen Kaplan and myself with numerous individuals over the past 20 years. Many of these were students who worked with us and have continued in this area of research subsequently. Others contributed by their searching questions. Janet Frey Talbot has been a colleague in this work for a very long time. In addition, we and I have been most fortunate to have the continuous support of the U.S. Forest Service through numerous Cooperative Agreements with the Urban Forestry Unit of the North Central Forest Experiment Station. Even more than the funding, we have cherished the encouragement and enthusiasm of the Unit's Project Leader, John F. Dwyer.

[2]Regression analysis shows that these three items account for almost half the variance ($R^2 = .49$) in explaining an index of Satisfaction with View (based on three other items). Beta coefficients are: $-.53$ for how difficult it is to see out, .24 for seeing natural elements, and .16 for likelihood of looking out.

## LITERATURE CITED

Balling, J. D. and J. M. Falk. 1982. Development of visual preference for natural elements. Environment and Behavior 14:5–28.

Frey, J. E. 1981. Preferences, satisfactions, and the physical environment of urban neighborhoods. Unpublished doctoral dissertation, University of Michigan.

Fried, M. 1982. Residential attachment: Sources of residential and community satisfaction. Journal of Social Issues 38:107–120.

Fried, M. 1984. The structure and significance of community satisfaction. Population and Behavior 7:61–86.

Kaplan, R. 1978. The green experience. In: S. Kaplan and R. Kaplan (eds.). Humanscape: Environments for people. (Republished in 1982 by Ulrich's, Ann Arbor, Michigan).

Kaplan, R. 1983. The role of nature in the urban context. In: I. Altman and J. F. Wohlwill (eds.). Behavior and the natural environment. Plenum, New York.

Kaplan, R. 1991. Environmental description and prediction: A conceptual analysis. In: T. Gärling and G. W. Evans (eds.). Environment, cognition, and action: An integrative multidisciplinary approach. Oxford University Press, New York.

Kaplan, R. and S. Kaplan. 1989. The experience of nature: A psychological perspective. Cambridge University Press, New York.

Kaplan, R. and J. F. Talbot. 1988. Ethnicity and preference for natural settings: A review and recent findings. Landscape and Urban Planning 15:107–117.

Kaplan, S., J. F. Talbot, and R. Kaplan. 1988. Coping with daily hassles: The impact of nearby nature on the work environment. Project Report. USDA Forest Service, North Central Forest Experiment Station, Urban Forestry Unit Cooperative Agreement 23-85-08.

Medina, A. Q. 1983. A visual assessment of children's and environmental educators' urban residential preference patterns. Unpublished doctoral dissertation, University of Michigan.

Moore, E. O. 1981. A prison environment's effect on health care service demands. Journal of Environmental Systems 11:17–34.

Talbot, J. F. and R. Kaplan. 1991. The benefits of nearby nature for elderly apartment residents. International Journal of Aging and Human Development. 33:119–130.

Ulrich, R. S. 1984. View through a window may influence recovery from surgery. Science 224:420–421.

Verderber, S. 1986. Dimensions of person-window transactions in the hospital environment. Environment and Behavior 18:450–466.

West, M. J. 1986. Landscape views and stress responses in the prison environment. Unpublished master's thesis, University of Washington.

Yang, B-E. and R. Kaplan. 1990. The perception of landscape style: A cross-cultural comparison. Landscape and Urban Planning 19:251–262.

CHAPTER 22

# The Restorative Environment: Nature and Human Experience

---

Stephen Kaplan

Professor of Psychology, University of Michigan

## INTRODUCTION

Early human beings were a part of nature. Over the millennia the gulf between humanity and the natural environment has steadily widened. Now, however, there is growing concern that this gulf has become too great, that we have strayed too far for our own good. This shift is due, at least in part, to a change in circumstances. Increasingly, people are confronted by pressures that are inexorably changing their lives. Although these pressures are by no means new, their steady increase and their cumulative impact are leading to increasingly unfortunate consequences.

Many of the pressures people face today are the results of three interacting forces: advances in technology, the knowledge explosion, and the increasing world population. Since these trends are more likely to get worse than better, they provide a useful working hypothesis as to what the pressures facing future populations might look like.

Although these trends each have distinct manifestations, they also have some common consequences. In particular, they all contribute to the experience of mental fatigue, which can lead people to be less tolerant, less effective, and less healthy. Natural environments can play a central role in reducing these unfortunate effects.

The thrust of my argument can summarized in terms of three basic themes:

1. Increasing pressures lead to problems of mental fatigue.

2. Restorative experiences are an important means of reducing mental fatigue, and have a special connection to natural environments.

3. Natural environments, in providing these deeply needed restorative experiences, play an essential role in human functioning.

These themes, in turn, lead to three groups of questions that I shall attempt to address:

1. The first set of questions concerns the pressures members of modern society face: Why are these pressures increasing? What impact do they have?

2. The second set concerns what Rachel Kaplan and I have come to call "restorative experiences," that is, experiences that help people recover from mental fatigue: What is the nature of these experiences? How do they achieve their substantial benefits? How does nature play a special role in providing such experiences?

3. Finally, what makes natural environments so important? What kinds of significant impacts can they have on the life of an individual?

## INCREASING PRESSURES AND THEIR IMPACTS

To understand the pressures facing an individual in the modern world, it is essential to understand something about the psychological process of attention, since attention is the aspect of human functioning that seems to suffer most.

For much of human history, information was scarce and the information available was highly selected (Postman, 1985). For a variety of reasons, the situation has changed dramatically in a relatively short time. Information is no longer scarce; as information proliferates, what is now scarce, as Herbert Simon (1978) has pointed out, is attention.

To appreciate what it means to say that attention is a scarce resource, a distinction that the great psychologist-philosopher William James made nearly 100 years ago may prove helpful. James (1892) identified a kind of attention, which he called "involuntary," that is evoked by something interesting or exciting in the environment. Such attention has the advantage of being effortless; attending to something of great interest is not hard work. Involuntary attention has two limitations. It is dependent upon an interesting environment, and sometimes one has to function in an environment that is not interesting. It also ties one to the environment; as such, it favors simple and direct responses rather than those that take advantage of one's higher mental processes.

A second kind of attention, which has come to be called "directed attention" (Stuss and Benson, 1986) does require effort. On the other hand, it permits one to focus selectively upon the environment, and to engage in higher mental processes such as problem-solving and planning. Unlike involuntary attention, directed attention is under voluntary control; when one instructs a child to "pay attention," one is referring to directed attention. The major limitation of directed attention is that it requires effort and that one's capacity to put forth that effort is finite. In other words, directed attention is susceptible to fatigue.

Given this brief sketch of the two kinds of attention, it is possible to begin to examine the pressures on attention that are characteristic of living in the modern world. Some of these pressures are the result of active competition for our attention. The mass media in general, and advertising in particular, are deeply committed to this informational struggle. High technology is employed to make these forms of informational competition ever more seductive. Thanks to marketing research and well-honed intuition, there is now considerable knowledge of what people find inherently interesting. This knowledge is effectively used against us, deflecting us toward stimuli that are hard to ignore but unsatisfying and unhelpful. Mander (1978) argues that in American television this coercive technology has been elevated to a highly refined art.

Some of the pressures on directed attention are not the outcome of an active struggle but take their toll nonetheless. It has become increasingly difficult to find the information one needs, embedded as it is in vast quantities of information that one does not need. Information retrieval has become so difficult that some corporations now favor doing a study on their own rather than searching the literature to determine whether it has been done before.

As the information explosion increases unabated and as media and advertisers fight over our scarce attention, the need for rest becomes increasingly important. Unfortunately, the trend has been in the opposite direction. The emphasis on efficiency and productivity, coupled with recent technological advances, has tended to reduce or eliminate the moments of rest that were at one time a natural part of everyday life.

The shift in the relationship of people to computers that has occurred in the past 30 years provides a vivid illustration. When I arrived in graduate school, the analysis of variance (a statistical technique) had recently been developed. With the aid of an electric calculator, such a statistic could be computed by a graduate student in about one year of work. This heroic effort then became the core of the student's dissertation. By the time I completed my graduate work, the computer had not only arrived on the scene but had become an accepted part of the institution's functioning. It was available to compute our statistics, and we had only to wait 24 hours for the results. Now waiting 24 hours for results would seem like an eternity; we expect to obtain them on-line, and many users are unhappy if their personal computers are not capable of multi-tasking, i.e., of doing several things at once. The rests one takes in between events are rapidly disappearing.

A story told by an Amish farmer provides a useful contrast (Kline, 1990). He was discussing the advantages of a horse over a tractor as a source of locomotive power on the farm. "Because God didn't create the horse with headlights, we don't work nights," he commented. Owning a tractor, by contrast, would lead to a powerful temptation to plow after dark. He also pointed out that the horses need to rest after a morning of work. As a result the family could assemble for lunch and a rest. Again, owning a tractor would have made it tempting to eliminate this important midday respite.

The increased pressure on directed attention forces us to expend more effort in order to retain focus on what is important. We thus call on directed attention with increasing frequency. At the same time, the decline in opportunities for rest leaves us less able to deal with the growing fatigue. The fatigue that results from these multiple assaults on our attention is not physical, but mental. Indeed, physical activity is often welcomed by individuals suffering from mental fatigue.

Everyone has experienced mental fatigue at one time or another; certainly everyone who has ever been a student remembers how one felt after completing final examinations. Despite this widespread experience, the implications and the seriousness of this mentally depleted state are not widely recognized. A compilation of the results of a variety of studies yields the following description of individuals suffering from this all-too-common condition: They have difficulty concentrating and are highly susceptible to distraction; find it difficult to make decisions; are impatient and inclined to make risky choices; are irritable and less likely than usual to help someone in distress; have difficulty either planning or carrying out previously made plans.

This is hardly a desirable state of affairs either for oneself or for someone with whom one associates. In extreme form, it could lead to excessive alcohol consumption or other drug abuse and/or to violent behavior. Even in milder form, such a state is unlikely to be conducive to creativity and effectiveness. Certainly if there were a way to reduce the overall level of mental fatigue in the population, it would be worth a substantial investment to do so. Fortunately, there is such a way.

## THE RESTORATIVE EXPERIENCE

The concept of restorative experiences arose in the context of a research program in the wilderness (Kaplan and Kaplan, 1989, chapter 4). The U.S. Forest Service had asked us to study the benefits of an ongoing wilderness program that was being carried out in Michigan's Upper Peninsula. Wilderness was not a primary research interest of ours, and the project was certainly not one we would have initiated on our own. What we learned from the research, however, was well worth the effort and turned out to have far broader applications than we would have expected.

The participants in the wilderness program we studied found the experience to be a profoundly restful and even healing one. In addition to recovery from mental fatigue, many of them found themselves in a reflective mode, stepping back to consider their lives and their priorities. They found nature more powerful, and at the same time more comforting, than they had ever imagined; they left the wilderness at the end of the trip worrying about how they could maintain their contact with this unexpectedly significant environment. An experience such as theirs, which leads to a recovery from mental fatigue as well as a variety of associated benefits, we have come to call a restorative experience.

Of particular interest for our present purposes are the four components of the restorative experience that we identified in the course of this research program. To understand these components and how they fit together, it might be helpful to pause for a moment to look at the thought process that occurs in restorative settings, and to see how it differs from the kind of thinking that occurs everyday. In this way, understanding the aspects of the environment that support this rather special pattern of thinking may be easier.

Let us return to the most basic aspect of the restorative experience, namely that it facilitates the recovery from mental fatigue. Our present task is to consider what goes on in the mind that accounts for this recovery process. This can be understood in terms of two basic themes:

1. Peoples' behavior depends upon the models of the world that they carry around in their heads.

2. When people can run that model effortlessly, they can rest that part of the mind that readily becomes fatigued.

Perhaps a little explanation is in order. What does it mean to say that one has a model of the world in one's head? Let us start with the assumption that under normal circumstances, people know something about what they are doing. Even a setting that one has never been in before may be sufficiently like other, familiar settings that one has some idea of what to do. In such cases, one has a model of the environment in one's head, and this model helps guide behavior (Kaplan and Kaplan, 1982).

Having a model is half the battle. To be able to run the model effortlessly, one also needs cooperation from the environment. Each of the four components of the restorative experience we identified in the wilderness research offers an essential aspect of this environmental support.

**Being Away.**    Being in some other setting makes it more likely that one can think of other things. People often talk of having to get away, of needing a change, when they are exasperated by the accumulation of mental fatigue (although they may not put it in those terms).

**Extent.**    Being away does not guarantee a restorative experience, however. Many settings may provide a change, but they are limited in scope. By contrast, restorative settings are often described as being "in a whole different world." Two properties are important to this experience: connectedness and scope; together they define what I mean by extent.

*Scope* requires that the environment is experienced as large enough that one can move around in it without having to be careful about going beyond the limits of the model that one is running. To have *connectedness*, the various parts of the environment must be perceived as belonging to a larger whole. Without that, one must repeatedly expend effort to find the model that is appropriate to the current momentary situation. A situation that allows a model to be left to run on "automatic pilot" requires far less effort.

Although the notion of extent is pertinent to a physical setting, it applies in a more conceptual, or imagined, sense as well. Thus, the experience of being in some distant "place" can also be realized when one is absorbed in a novel or by a performance.

**Fascination.**    In addition to the need for extent, restorative experiences depend upon interest or fascination. A fascinating stimulus is one that calls forth involuntary attention. Thus fascination is important to the restorative experience, not only because of its intrinsic attraction, but also because fascination allows one to function without using directed attention. Here, too, the ease with which one can run one's mental model of the world is directly affected. Without fascination, there is always the danger that the model one should be running will give way to distraction or to daydreaming. Effort is required to keep the appropriate model in focus. One of the great benefits of fascination is that it frees one from the need for effort of this kind.

Just as extent can be based on the physical environment or on one's perceptions and thoughts, fascination can be derived from objects in the environment, as well as from ways of doing things. People are fascinated by figuring things out, by predicting uncertain events, by challenges. Thus, restorative experiences can draw on a great variety of circumstances, as long as there is sufficient extent and enough to keep one absorbed by it.

**Compatibility.**    Even with fascination and extent, an environment can still fall short as a setting for restorative experiences. The final component of the restorative concept calls upon the compatibility among the environmental patterns, the individual's inclinations, and the actions required by the environment (S. Kaplan, 1983).

The importance of compatibility is easiest to see in its absence. There is no lack of settings in which the environment undermines what one is trying to accomplish, where one's goals and actions are obstructed by the demands made by the environment. Such situations require considerable mental effort. In a compatible environment, by contrast, what one wants to do and is inclined to do are what is needed in and supported by the environment. When what intuitively feels right is what the situation requires, one's model is thoroughly supported by what is happening in the environment. In such cases, one's relationship to the environment takes on an effortless quality that can be deeply restorative.

Although these properties of a restorative experience emerged in the context of the wilderness experience, it quickly became evident that they were by no means unique to such settings. In particular, the garden experience (R. Kaplan, 1973), which is different in so many ways, turned out, on a deeper level, to have striking similarities.

## APPLYING THE RESTORATIVE CONCEPT TO THE NATURAL ENVIRONMENT

Although the restorative environment is by no means restricted to natural settings, natural environments seem to be particularly restorative. Of particular importance in this context is the role of "accessible nature." This role should become increasingly clear as we examine each of the properties of the restorative experience discussed above in the context of natural settings in general. The emphasis in this section is on settings that, while often undramatic and small in scale, have the essential property of being readily accessible.

**Being Away.**    Natural settings are often the preferred destinations for extended restorative opportunities. The seaside, the mountains, lakes, streams, forests, and meadows are all idyllic places for "getting away." Yet for many people in the urban context, opportunities for getting away to nature spots in their nearby environment are minimal. Natural environments that are easily accessible thus offer an important resource for resting one's directed attention.

**Extent.**   In the distant wilderness, extent comes easily. But extent need not entail large tracts of land. Even a relatively small area can provide a feeling of extent. Trails and paths can be arranged so that a small area seems much greater. Miniaturization provides another device for providing a feeling of being in a whole different world, though the area is in itself not extensive. Japanese gardens sometimes combine both of these devices in giving the sense of scope as well as connectedness.

Extent, as already mentioned, also functions at a more conceptual level. For example, settings that include historic artifacts can promote a sense of being connected to past eras and past environments and thus to a larger world.

**Fascination.**   Nature is certainly well-endowed with fascinating objects, as well as offering many processes that people find engrossing. Many of the fascinations afforded by the natural setting might be called "soft fascination." Clouds, sunsets, snow patterns, the motion of the leaves in a breeze—these readily hold the attention, but in an undramatic fashion. Attending to these patterns is effortless, and they leave ample opportunity for thinking about other things.

When one thinks of sources of soft fascination, vegetation is a recurring theme—the view of trees and grass out the window, masses of flowers, the garden. People find these patterns aesthetic and pleasurable; in the context of this pleasure, people can reflect on difficult matters that would be too confusing or too painful to contemplate under other circumstances.

**Compatibility.**   The natural environment is experienced as particularly high in compatibility. It is as if there were a special resonance between the natural setting and human inclinations. For many people, functioning in the natural setting seems to require less effort than functioning in more "civilized" settings, even though they have much greater familiarity with the latter.

It is interesting to consider the many patterns of relating to the natural setting. There is the predator role (such as hunting and fishing), the locomotion role (hiking, boating), the domestication of the wild role (gardening, caring for pets), the observation of other animals (bird watching, visiting zoos), survival skills (fire building, constructing shelter), and so on. People often approach natural areas with the purposes that these areas readily fulfill already in mind, thus increasing compatibility.

A nearby, highly accessible natural environment cannot provide the context for all of these goals and purposes. Yet even such a setting is likely to be supportive of the inclinations of those who seek a respite there. It is amusing to think of the factory worker who races off during the lunch period, fighting traffic and distractions, to find a spot in the shade of a tree for a peaceful break. If the peaceful effect would have been totally worn off by the time the return trip is made at the end of the hour, would this ritual be repeated again tomorrow?

## CONCERNING THE IMPORTANCE OF THE NATURAL ENVIRONMENT: RESEARCH RESULTS

A recently completed study (Cimprich, 1990) brings together many of the themes of the conference in a particularly striking fashion. Years of experience working as an oncology nurse at Sloan-Kettering Institute in New York led Dr. Cimprich to wonder why preparing patients to care for themselves after they left the hospital was so difficult. She also wondered why patients who no longer required treatment and were considered to have excellent prospects from a medical point of view, so often experienced serious problems. They reported difficulties in managing their lives and difficulties with their marriages. They experienced symptoms they were unable to explain, which disturbed them and led them to distrust the clean bill of health they had been given. One of the characteristic difficulties, involving severe limitations in the capacity to focus, has been found to persist for at least three years after treatment. Patients often compensate for this limitation by a progressive narrowing of their lives.

A number of years ago, Dr. Cimprich, then a doctoral student in the School of Nursing at the University of Michigan, was taking a course with me. When the topic of directed attention and mental fatigue was discussed in class, she immediately recognized its relevance to the plight of the cancer patients with whom she had worked. When viewed from this perspective, the experience of the cancer patient can be seen as a multifaceted attack on the unfortunate individual's directed attention.

For her dissertation project, Dr. Cimprich decided to work with breast cancer patients who had an excellent prognosis for a full recovery. She set out to find a battery of objective measures of attentional capacity. Lacking any pure measures of directed attention, I proposed that we add a novel technique to the battery of tests. Many people are familiar with the wire frame drawing of a cube (technically called a "Necker cube" in the psychological literature) that appears to "reverse" as one focuses upon its center for a period of 30 seconds. In other words, which face of the cube appears to be in front seems to change from time to time, assuming that one stares at it long enough. The rate of these reversals varies considerably from one person to the next.

Although the basic phenomenon of the apparent cube reversal is widely known, what is less well known is that it is possible to slow the rate of reversal by making an effort to do so. If one focuses one's attention on the cube as it appears at a given moment, its apparent change to the other possible cube will in general be delayed. This phenomenon is of great theoretical interest because directed attention is believed to achieve its focusing influence by inhibiting potentially interfering material. Thus by asking the participants in this study to attempt to slow the rate of reversal of the cube, we could assess the strength of this inhibitory factor relatively directly.

Participants in this study were randomly assigned to one of two groups. Individuals in the intervention group were told that people in their condition sometimes benefit from setting aside some time for restorative experiences. These were explained and a list of potentially restorative activities was offered. From this list they chose four activities. These were then listed in a contract that stated the patient's commitment to participate in at least three such activities per week, with a minimum duration of 20 to 30 minutes. Both the patient and Dr. Cimprich signed the contract.

Each patient was tested four times on the battery of attentional measures (over a period of 90 days since surgery). At the initial testing point, the average scores of these patients were so low as to fall in the range characteristic of brain-damaged patients. The recovery of the control group was erratic and uneven; many of their scores at the end of the period of testing were not significantly different than at the beginning. By contrast, the intervention group showed improvement on all measures. Perhaps most striking was their performance on the "slowed" Necker Cube task. It showed steady improvement throughout the testing interval.

The benefits of this remarkably modest intervention were not restricted, however, to measures of attention. There were two areas in which there were indications that the intervention group was well on its way to returning to a normal, healthy life. More of these participants went back to work during these initial 90 days, and more of them went back full-time. Perhaps most fascinating of all, many of the participants in the intervention group started new projects during this time. They decided to undertake challenges such as losing weight or learning a new language. None of the participants in the control group could think of any new projects they had initiated during that interval.

No single study can in itself be considered definitive. Replications and extensions of this important work are clearly needed. Yet the insights and implications of this initial study are too fascinating to ignore. Let us examine a few of them:

**1.** Out of the various restorative activities on the list, the predominant choice by far was of nature-related activities.

**2.** It is unlikely that breast cancer patients are the only ones who suffer assaults on their attention or could benefit from systematic participation in restorative activities. Many illnesses, traumatic experiences, and difficult life transitions can be seen in this new context as likely to place extreme demands on directed attention and to call for similar treatment. The limited attentional capacities of the elderly suggest that they, too, could benefit. It is interesting to contemplate the pressure on natural settings, the increased demand for garden opportunities, and the other claims on this limited resource that would occur if all groups who could benefit knew about and partook of this opportunity. Either we need to keep the information a well-guarded secret, or we need to redouble our efforts to preserve and conserve open-space.

**3.** We are now a step closer to understanding the effect of the natural environment on human health. The impressive studies of improved health in prisons (Moore, 1981; West, 1986) and enhanced healing in hospitals (Ulrich, 1984; Verderber, 1986) when a more restorative view was available have made it abundantly clear that such an important relationship existed. It is now possible to begin to grasp one of the underlying mechanisms. There is good reason to believe that activities essential to maintaining, or, indeed, recapturing one's quality of life are dependent on directed attention. Further, it appears that a fatigued directed attention is benefitted by even a modest amount of restorative activity.

**4.** The difference between nature as an amenity and nature as a human need is underscored by this research. People often say they like nature; yet they often fail to realize that they need it. The same restorative opportunities were available to all participants in this study; there is every reason to believe that they all would have benefitted. Yet only the participants who had contracted to do so consistently carried out these activities. An educational and a cultural gap needs to be addressed. Nature is not merely "nice." It is not just a matter of improving one's mood, rather it is a vital ingredient in healthy human functioning.

## ACKNOWLEDGMENT

Work on this paper and on the research that led to it was supported, in part, by the U.S. Forest Service, North Central Forest Experiment Station, Urban Forestry Project, through several cooperative agreements. I would also like to thank Rachel Kaplan for her many and substantial contributions to this paper.

## LITERATURE CITED

Cimprich, B. 1990. Attentional fatigue and restoration in individuals with cancer. Doctoral dissertation, University of Michigan.

James, W. 1892. Psychology: The briefer course. Holt, New York.

Kaplan, R. 1973. Some psychological benefits of gardening. Environment and Behavior 5:145–152.

Kaplan, R. and S. Kaplan. 1989. The experience of nature: A psychological perspective. Cambridge University Press, New York.

Kaplan, S. 1983. A model of person-environment compatibility. Environment and Behavior 15:311–332.

Kaplan, S. and R. Kaplan. 1982/1989. Cognition and environment: Functioning in an uncertain world. Praeger, New York. (Republished in 1989 by Ulrich's, Ann Arbor, Michigan)

Kline, D. 1990. Great possessions: An Amish farmer's journal. North Point Press, San Francisco.

Mander, J. 1978. Four arguments for the elimination of television. Morrow-Quill, New York.

Moore, E. O. 1981. A prison environment's effect on health care service demands. Journal of Environmental Systems 11:17–34.

Postman, N. 1985. Amusing ourselves to death. Penguin Books, New York.

Simon, H. A. 1978. Rationality as process and as product of thought. American Economic Review 68:1–16.

Stuss, D. T. and D. F. Benson. 1986. The frontal lobes. Raven, New York

Ulrich, R. S. 1984. View through a window may influence recovery from surgery. Science 224:420–421.

Verderber, S. 1986. Dimensions of person-window transactions in the hospital environment. Environment and Behavior 18:450–466.

West, M. J. 1986. Landscape views and stress responses in the prison environment. Unpublished master's thesis, University of Washington.

CHAPTER 23

# Vegetation and Stress:
# A Comparison Study of Varying Amounts of Vegetation in Countryside and Urban Scenes

Mary Krehbiel Honeyman

Landscape Architect, Oblinger, Mason, McCluggage and VanSickle, Corp.

## EXPERIMENTAL METHODS

The effect of the visual perception of varying amounts of green vegetation on human stress levels was the central focus of this research study. Pre- and post-tests of stress were measured with the Zuckerman Inventory of Personal Reactions (ZIPERS), a ten-question broad-affect test that measures an individual's emotions and anxiety state at the moment it is given. The test assessed five feelings: fear, anger, positive affect, sadness, and attentiveness. A demographic questionnaire was used to collect background information. The procedure involved showing colored slides of outdoor environments to three groups of participants. A total of 213 college students participated in the experiment immediately following hour-long examinations. The participants were randomly divided into three groups of nearly equal size: *countryside, urban with vegetation, urban* (without vegetation). Participants then completed the ZIPERS test. Following the stress test, the three slide presentations were given. Slides in all presentations were shown at the same intervals of 9 to 10 seconds, creating a total presentation time of 5 to 6 minutes. The *countryside* group was shown slides of green, vegetated countryside. The *urban with vegetation* group viewed slides showing urban scenes with vegetation introduced. The *urban* group viewed slides of urban scenes without vegetation. Upon completion of the slide presentations, the participants were asked to complete a second ZIPERS test identical to the first. Following the second test, the participants completed a questionnaire that requested demographic information.

## RESULTS

Data collected were then sorted and analyzed. A series of t-tests were used to compare pre- and post-test responses within each of the three groups. Results of the t-tests revealed significant differences in the participants' responses to the ZIPERS questions in four of the affect factors as shown in Table 1, with probability set at .05 or less.

When one looks closely at the *P*-values, the decline f the affect factor of "positive affect" in the *urban* group with a *P*-value of 0.0001 shows, by far, the most dramatic significant difference of the four. The decline in fear for *countryside* and both fear and anger for *urban with vegetation* are also significant.

These results indicate a trend toward a reduction in stress for the groups that viewed scenes with vegetation included, and an elevation in stress levels for the group that viewed scenes with no vegetation included. Prior to data analysis, it was hypothesized that the participants' stress levels would be reduced as the amount of vegetation in the scenes was increased, with the *urban* group showing the least stress reduction and the *countryside* group showing the most stress reduction. Results of the data analysis revealed apparent trends toward stress reduction in both of the groups that viewed scenes including vegetation, and trends toward stress elevation in the *urban* group, which viewed scenes with no vegetation. The *countryside* group showed trends toward reduction of stress in three of the five affect factors (fear, anger, and positive affect) and elevation of stress in two of the five affect factors (sadness and attentiveness). The *urban* group showed trends toward elevation of stress in four of the five affect factors (anger, sadness, positive affect, and attentiveness) and a reduction of stress in the affect factor of fear. To our surprise, the *urban with vegetation* group showed trends toward reduction of stress in all five of the affect factors. These trends seemed to indicate that the inclusion of vegetation in the urban environment, rather than simply the amount of vegetation in the scenes, appeared to have the greatest psychological impact on the observer.

**Table 1.** Differences in affect scores before and after slide presentations.

| Affect factor | Mean difference | *P*-value |
|---|---|---|
| COUNTRYSIDE GROUP | | |
| Fear | −0.111 | 0.0039[a] |
| Anger | −0.134 | 0.0903 |
| Sadness | 0.028 | 0.6406 |
| Positive affect | 0.106 | 0.1454 |
| Attentiveness | −0.194 | 0.1366 |
| URBAN GROUP | | |
| Fear | −0.071 | 0.1989 |
| Anger | 0.033 | 0.6064 |
| Sadness | 0.129 | 0.2007 |
| Positive affect | −0.475 | 0.0001[a] |
| Attentiveness | −0.171 | 0.1529 |
| URBAN WITH VEGETATION GROUP | | |
| Fear | −0.169 | 0.0039[a] |
| Anger | −0.155 | 0.0083[a] |
| Sadness | −0.056 | 0.5088 |
| Positive affect | 0.039 | 0.5817 |
| Attentiveness | 0.086 | 0.4894 |

[a]Significant Difference—Probability .05 or less.

## CONCLUSION

From this study, one can infer the following:

1.  The natural environmental element of vegetation plays a role in human psychological well-being.

2.  The inclusion of vegetation in urban areas provides psychological benefit to those who experience it.

3.  The exclusion of vegetation in urban areas actually elicits negative psychological human responses, thereby creating stress.

4.  Human psychological response is impacted by the absence or presence of vegetation in the urban context.

5.  The introduction of vegetation in the urban environment has a positive psychological impact upon those who experience that landscape.

The results of this experiment support the inclusion of vegetation and green space in urban design to contribute to the psychological well-being of the general public. Information of this type also provides evidence in support of the work of landscape architects who include vegetation in urban design and, thereby, create psychologically beneficial urban spaces. As an important moral and civic responsibility, all urban designers should carefully consider the psychological impact that a designed space may have on the public.

CHAPTER 24

# *What We May Learn Through Horticultural Activity*

Eisuke Matsuo

Faculty of Agriculture, Kagoshima University, Japan

## INTRODUCTION

According to the report on leisure activity in Japan (Yokakaihatsu Center, 1988), 32.2 million people, or approximately 32% of the Japanese population, enjoy horticulture as a leisure activity or hobby. Horticulture holds first place among the hobby and leisure activities in Japan. Why does horticulture as a hobby or leisure activity fascinate us? In the United States and Great Britain, gardening is a common treatment for mentally retarded or physically handicapped individuals (Relf, 1981). Why should such activity provide effective treatment? As the number of elderly people increases in Japan, providing for their welfare is of increasing concern. Horticulture is one of the effective means for enchancing the quality of life of the elderly. Why can horticulture help keep the elderly's lives worth living? The answer to these questions is that horticultural activity makes us feel that we are human beings.

## PEOPLE HAVE TWO WAYS OF BEING CREATIVE

Animals are instinctive, but people are creative (Ichikawa, 1970; Tokizane, 1974). When we are creative, we have a sense of relatedness and even pleasure. People have two ways of being creative: by fostering life and by acquiring objects (Matsuo, 1982). These activities are based on different philosophies: the philosophy of fostering and that of acquiring. The philosophy of fostering originated in the maintenance of the human race and that of acquiring in the maintenance of the individual body. That is, *Homo sapiens* evolved from an animal, instinctive being into a human, creative being, and developed through the philosophies of fostering life and of acquiring objects. These two philosophies and their related activities make it possible for us to live as human beings.

Concrete examples of fostering life are growing plants, raising animals, educating children, or training successors. Examples of acquiring objects are making things, obtaining material and information, manufacturing, fishing, hunting, shopping, reading, writing, appreciating arts, etc.

There are important differences between these two ways of being creative: (1) By fostering life, although we only help life to develop according to its genetic information, we give ourselves to it, and we often feel or find ourselves a part of it. It is necessary to keep company with that which we foster during its growth for long time. Our attitude toward these living things becomes nurturing and supportive. (2) By acquiring objects, we accomplish our aims by our own will. Our attitude toward objects is discriminating, volitional, i.e. we take and choose.

Because people evolved to be creative in both ways, we live as human beings by responding to the urge to foster and the urge to acquire. Therefore, the following can be concluded: (1) If we discard one or both of these urges, we abandon living as a human being. (2) If we are unable to satisfy either urge, we are prevented from living as a human being.

## HORTICULTURE AND HORT-INDUSTRY

If we examine our daily life, we may be surprised to realize that it is full of the philosophy and behavior of acquiring objects but lacking in that of fostering life. Even farming, which many consider imbued with the philosophy of fostering life, is directed toward economical and efficient production. Today, farming is carried out with the concept and behavior of acquiring. Horticulture concerned with production is done primarily on a commercial basis. The goal is to obtain a profit by producing and marketing horticultural products. For this purpose, mechanization, specialization, and mass-production have been promoted, thereby introducing the principle of industry into horticultural production. This attitude is quite acquisitive, and reveals the concept and behavior of acquiring that is typical in common horticulture. Therefore, this type of horticulture should be referred to as "hort-industry," which means horticulture based on an economic basis, similar to the term "agri-business."

In contrast to hort-industry, horticulture done for pleasure or for enhancing the quality of human life is practiced mainly by amateur horticulturists who do not primarily target profits from horticultural products as their goal, but take care of their plants as if they were children. Growing Bonsai or some other potted plants, gardening or culturing Volkstuin in Holland, Kleingarten in Germany, Allotment in England, and Shimin-Noen in Japan, may be good examples of this. This kind of horticulture is conducted according to the philosophy and behavior of fostering life. Therefore, it could be termed the real "horticulture," because it is a part of life and culture. It is this meaning of the term "horticulture" that is used in the discussion below.

## TWO WAYS OF ACTING CREATIVELY THROUGH HORTICULTURE

When we ponder our daily horticultural activities, we notice many interesting experiences. For example, we are pleased when our plants grow well day by day; we are disappointed when plants die off or become damaged for some reason or wilt due to drought because we forgot to water; having given all necessary care to our plants over a long period, we enjoy their subsequent flowering and fruit setting; we do not want to sell our dear plants, even if someone is prepared to pay any price for them. In many ways, we feel toward our plants as we do toward our children. These examples show the spirit behind the fostering of plants.

In addition, we also enjoy the beauty of flowers and the harvesting and eating of fruits and vegetables. In Japan, a popular pastime is to tour the orchards to pick fruits such as apples, pears, mandarins, strawberries, or grapes, and to go into the fields to dig up sweet potatoes, peanuts, or potatoes. These are typical examples of our philosophy behind acquisitive horticulture.

Thus, horticulture provides us not only with the concept and behavior of fostering life through taking care of plants, but also with the concept and behavior of acquiring objects through harvesting and/or admiring horticultural products and accomplishments that are obtained by our own effort. In other words, horticulture fulfills the two ways of being creative, whereas most other hobbies do not because they are generally only acquisitive, and the fostering of life is available only through raising living things. Therefore, horticulture differs from other hobbies in that it enables us to live as human beings. This is also why it is used as a therapeutic method e.g. Horticultural therapy, why it has become a hobby by which we are fascinated, and why it is one method for keeping the elderlys' lives worth living.

Too much emphasis on economic demands is not good for living our lives as full human beings or for horticulture. Modern life, particularly in the big cities, prevents people from fostering life. It may therefore be seen as natural that people are eager to foster plants and thereby to live as more fully human. That is the reason horticulture has become very popular.

## ACKNOWLEDGMENT

The author thanks Dr. J. Latimer, University of Georgia, for reading this manuscript.

## LITERATURE CITED

Ichikawa, K. 1970. Souzou ni ikiru ningen-Homo creata. Koudansha, Tokyo. (in Japanese).
Matsuo, E. 1982. Why horticulture is necessary for human life. Alps Insatsu, Kagoshima. (in Japanese).
Relf, P. D. 1981. Therapy and rehabilitation through horticulture. Chronica Horticulturae 21(1):1–2.
Tokizane, T. 1974. Ningen de arukoto. Iwanami Shinsho, Tokyo. (in Japanese).
Yokakaihatsu Center. 1984, 1988. Leisure Hakusho '84, '88. Yokakaihatsu Center, Tokyo. (in Japanese).

CHAPTER 25

# Urban Nature, Place Attachment, Health, and Well-Being

R. Bruce Hull, IV

Associate Professor of Architecture, Texas A & M University

Gabriela Vigo

Graduate Research Assistant, Landscape Architecture, Texas A & M University

The concept of place-attachment seems an appropriate focus for environmental planners and designers concerned with urban nature. Place-attachment is the "glue" that binds people to place (Rivlin, 1982). It is the bond between residents and their community (Stokols and Shumaker, 1982; Fisher, 1977). The task ahead is to demonstrate that urban vegetation influences place-attachment. But first, it is appropriate to at least partially document that place-attachment is relevant. A review of the literature suggests that place-attachment provides benefits to users of a place that outweigh the costs of promoting it. Place-attachment impacts socially significant, economically tangible factors such as social interaction (Unger and Wandersman, 1985), sense of coherence and health (Antonovsky, 1987; Freeman, 1984), positive moods such as belonging and pride (Stokols and Shumaker, 1982), public involvement and self governance (Unger and Wandersman, 1985), residential mobility (Stokols and Shumaker, 1982), altruistic behaviors, stress, property values, crime, vandalism, and so on. The list and documentation of impacts are impressive and still growing. Space limitations do not allow full treatment here, however. Suffice it to say that place-attachment seems a worthy objective of planners, designers, politicians, and others involved in determining the form and quality of our environment.

Place and attachment to place are promoted by layering opportunities, meanings, and emotions on a setting (Figure 1). The denser and more interrelated the layers, the more likely the setting will develop the qualities of place. There are many types of layers, but we discuss only a few here. Urban nature can play a significant role in each of these layers.

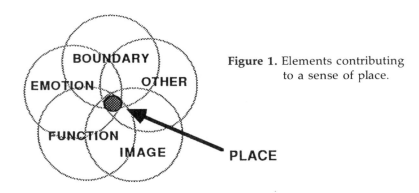

**Figure 1.** Elements contributing to a sense of place.

*Boundaries*, boundary themes, and landmarks differentiate one place from another; they denote that there is a there, there. Obvious boundaries promote feelings of membership or increase awareness that one is trespassing, depending upon whether one belongs in the setting or not. These qualities of place promote sense of community and, consequently, place-attachment (McMillan and Chavis, 1986). Trees and other planting materials are part of the landscape and urban designers' palette. They are a powerful and frequently used means of delineating space (i.e., boundaries) and providing distinctive character to places (i.e., boundary themes). Trees and parks also serve as community landmarks.

*Image expression* refers to the act of using physical features that evoke salient meanings understood by others in the community to communicate information about the values, concerns, priorities, activities, and socioeconomic status of one's self, i.e., to express the extended-self (Belk, 1988). Hair and dress styles are perhaps the most commonly exploited means of expressing one's image. Automobiles, houses, yards, and even communities serve similar roles. In addition to communicating information about one's self through characteristics of place, image expression provides an important opportunity to test others' reactions to one's continually evolving image of self. *Image testing* allows one to try images "on for size" and assess others' reactions without making the inner-self vulnerable. Similarly, *image comparison* occurs when one compares the values and concerns of others, as symbolized by their environment, with one's own values and concerns. Whether or not the values symbolized are congruent with one's image of self, comparison opportunities may help better define the self. *Image congruity* is how well a place's image corresponds to the individual's image of self. Congruity promotes place-attachment. Opportunities for image expression and testing of self are important because they are vital links to health and well-being. They provide opportunities to strengthen and promote a sense of self-worth and ultimately a sense of coherence and health. These concepts are closely linked to the concept of place identity (Korpela, 1989; Proshansky et al., 1983).

Vegetation in residential areas provides excellent opportunities for image expression and testing. Vegetation of all types, ranging from lawns to flower boxes, are powerful symbols. Vegetation is easy and relatively inexpensive to manipulate. Vegetation also readily evidences effort—a mulched flower bed or a trimmed yard immediately documents the labor input. Studies have documented that people regularly associate meanings with the type and maintenance of vegetation in front yards: conservative or liberal, family or single, seclusion and privacy, low versus high social status (Rapoport, 1982; Hull, 1990) are but a few examples.

An additional layer contributing to place attachment is functional congruity. *Functional congruity* is how well a place meets the functional needs of the user; i.e., the classic problem of person-environment fit. In residential areas, the fit depends largely on the life style and life stage of the resident (Michelson, 1976). For example, some may need access to play areas and community activity centers whereas others may need exciting social opportunities or opportunities for privacy and relaxation; all may require safety. A related concept, *place dependency*, reflects the ease with which one can "pick up roots" and transport them to a new location. "The place dependence of settings is defined as the degree to which group members perceive the major functions of their settings to be exclusive to a particular location" (Stokols, 1981). Strong place-attachment results when one is highly place dependent and one's needs are met by the available services and opportunities. Both conditions—place-dependency and functional congruity—are required for strong place attachment.

Urban nature promotes functional congruity and place-dependency in numerous ways. Contact with nature is associated with relaxing and restorative experiences (Kaplan and Kaplan, 1989; Hull and Harvey, 1989). These opportunities are highly sought after by many people. Nature also provides a context for many leisure pursuits. It provides shade and thus facilitates outdoor activities in the heat of summer. Visual access to nature is associated with aesthetics and property value. These and other factors facilitate functional congruity. Past development philosophies have contributed to making the abundance of vegetation a scarcity in residential areas and thus one of those qualities difficult to replace if relocation were required. Thus, if one lived in a place where access to nature was plentiful, one might become dependent upon that resource, because replacing it in other locations would be difficult.

Another characteristic that differentiates place from space is the *feelings* associated with place. When one enters a place, one feels something. Emotions felt in a place are another important layer that contributes to place and place-attachment. As mentioned above, contact, even visual contact, with nature promotes relaxed feelings. Nature is also a means of providing opportunities for prospect-refuge in public place; this promotes feelings of relaxation, seclusion, and safety. Thus nature by its presence or absence can be used to elicit affect in place users.

Findings from empirical studies in communities in Australia and the United States support the proposed conceptual model. Image comparison and image expression with nature play important roles in peoples' attachment to place. Arbor Day, commemorative tree planting when a child is born, and the growing popularity of street tree programs evidences the value placed on this type of vegetation. Moreover, access to nature is a valued and sought-after resource in many communities; as is evidenced by studies of persons' favorite places in their residential areas, more often than not the preferred places are heavily vegetated. The desire to experience nature is also evidenced in the vast literature on outdoor recreation use. Finally, there is mounting evidence that nature has a powerful impact on peoples' emotional states. These factors combine to suggest that nature can be effectively used to promote place-attachment and all the benefits derived from it.

## *LITERATURE CITED*

Antonovsky, A. 1987. Unraveling the mystery of health: How people manage stress and stay well. Jossey-Bass, San Francisco.

Belk, R. W. 1988. Possessions and the extended self. Journal of Consumer Research 15:139–168.

Fisher, C. S. 1977. Networks and places. Free Press, New York.

Freeman, H. 1984. Housing. In: Mental health and the environment. H. Freeman (ed.). Churchill Livingstone, London.

Hull, R. B. 1992. Incongruity, place attachment and community design. Journal of Architectural and Planning Research. 9:1 (in press).

Hull, R. B. and A. Harvey. 1989. Explaining the emotion persons experience in suburban parks. Environment and Behavior 21:323–345.

Kaplan, R. and S. Kaplan. 1989. The experience of nature: A psychological perspective. Cambridge Univ. Press, Cambridge.

Korpela, K. M. 1989. Place-identity as a product of environmental self-regulation. Journal of Environmental Psychology 9:241–256.

McMillan, D. W. and D. Chavis. 1986. Sense of community: A definition and theory. Journal of Community Psychology.

Michelson, W. 1976. Man and his urban environment: a sociological approach. Addison Wesley, Reading, MA.

Proshansky, H. M., A. K. Fabian, and R. Kaminoff. 1983. Place identity: Physical world socialization of the self. Journal of Environmental Psychology 3:57–83.

Rapoport, A. 1982. The meaning of the built environment: A nonverbal communication approach. Sage, Beverly Hills.

Rivlin, L. 1982. Group membership and place meanings in an urban neighborhood. Journal of Social Issues 38:75–93.

Stokols, D. 1981. Group × place transactions: Some neglected issues in psychological research on settings. pp. 393–415. In: D. Magnusson (ed.). Towards a psychology of situations: An interactional perspective. Lawrence Erlbaum, Hillsdale, New Jersey.

Stokols, D. and S. A. Shumaker. 1982. The psychological context of residential mobility and well-being. Journal of Social Issues 38:149–171.

Unger, D. G. and A. Wandersman. 1985. The importance of neighbors: The social, cognitive, and affective components of neighboring. American Journal of Community Psychology 13:139–169.

CHAPTER 26

# Regional Connectedness: Urban, Rural, and Primeval

Robert G. Dyck

Professor of Urban Affairs and Planning,
Virginia Polytechnic Institute and State University

Magne Bruun

Professor of Landscape Architecture, Agricultural University of Norway

Anupam Mukherjee

Graduate Program in Urban/Regional Planning,
Virginia Polytechnic Institute and State University

## INTRODUCTION

This analysis develops a perspective on strategies for connectedness in regional development that has international applicability, even though it is based in large part on ideas developed by selected U.S. authors. Following a brief review and critique of these ideas, we present a short list of developmental policies and models applicable in Virginia, together with a case study of a semirural area in western Virginia and a list of research activities needed for improved settlement policy in Virginia.

## SUMMARY OF SELECTED REGIONAL CONNECTEDNESS IDEAS: QUESTIONS POSED FOR THIS STUDY

**MacKaye (1928).** Every citizen should have access to experiencing the "connectedness" of urban, rural, and primeval landscapes as a basis for positive health and the perception of personal and social well-being. Our questions:

1. The connectedness argument is basically intuitive. Can it be demonstrated on the basis of specific scientific analysis?

2. Can the urban environment be improved so as to reduce use pressure by urban residents on the rural and primeval environment?

**Yaro (1988).** Working rural landscapes should be protected from typical suburban and exurban development by design regulation and high development densities to assure affordable housing and to preserve agricultural productivity, historical values, and ecological values. Our questions:

1. Does this strategy deal adequately with urban and primeval landscapes?

2. Does it leave adequate space for additional rural development in the future? Does it adequately protect the agricultural landscape?

3. Does it consider adequately the imperatives of energy conservation in settlement design and commutation, as related to pending depletion of non-renewable energy sources?

**Lewis (1979a,b).** Community plantscapes and gardens have a demonstrably positive impact on the physical, psychological, and social well-being of low-income urban dwellers. The importance of trees and gardens in the urban landscape, often denigrated by functionalist designers, is thus re-established. Our questions:

1. Has Lewis succeeded in relating the micro-environment to the larger townscape and regionscape (macro-environment)?

2. Is the focus on private gardening sufficient to solve the larger social need for community treescapes?

**Brown (1990).** Environmental degradation has reached crisis proportions worldwide. The trends of global warming, tropical deforestation, depletion of top soil, etc., must be reversed by the year 2000. New kinds of sustainable economic and living environments must be established by the year 2030. Urban environments may need to be decentralized. Walking and bicycles will have to replace universal reliance on fuel-wasteful automobiles. Our questions:

1. Aren't *more* rather than less concentrated urban environments more compatible with walking and bicycling?

2. Can existing global trends towards urban concentration be reversed in 10 or even 40 years?

## APPLICABLE DEVELOPMENT POLICIES AND MODELS

### State and Regional Level Policies and Programs in Virginia

The Virginia Growth Management Forum (September 1989) recommended the following:

1. Establish a growth management structure and plan for the Commonwealth.

2. Enhance local government ability to cope with growth pressures.

3. Improve regional planning capabilities.

4. Empower the Virginia Commission on Population Growth and Development (subsequently approved by the 1990 General Assembly).

The 1989 Virginia Outdoors Plan (Commonwealth of Virginia, Department of Recreation and Conservation, 1989) shows that participation in outdoor recreational activities is increasing at a faster rate than the population as a whole, and that the greatest demands are for close-to-home recreational areas and facilities for activities such as hiking, bicycling, etc. Open space needs for 12 state regions are projected to the year 2000, including demand from populations outside the respective regions.

## Models

The Massachusetts plan for protection of characteristic working rural landscapes and sustainable development provides an approach to resolving tensions between the private real estate market and sustainable resource utilization (Yaro, 1988). It requires new approaches to rural subdivision and zoning regulation designed to increase housing and business development density while preserving significant open spaces. New kinds of professional leadership related to local citizens' participation are also indicated.

The Massachusetts approach requires a closer linkage of state, regional, and local planning; it is no longer viable to separate state, regional, and local settlement planning decisions. Note that this approach is consistent with MacKaye's intuitive recognition of the necessary linkages among the urban, rural, and primeval landscapes; connectedness cannot be achieved by dealing separately with each category. We explore the meaning of this approach in the context of the following illustrative case study of the Price's Fork area near Blacksburg, Virginia.

## PRICE'S FORK CASE STUDY

### Parkway Plan

The Blacksburg Town Planning Commission has prepared a corridor plan that defines the area of concern (Price's Fork Road Corridor Study Committe, 1985). The area is a scenic strip approximately 1.5 miles wide extending between Price's Mountain and Brush Mountain and between Blacksburg and the village of Price's Fork. Planning imperatives drawn from the existing plan and this study are the following:

1. Protection of existing working farms.

2. Protection of significant viewsheds.

3. Protection of woodlands and water courses.

4. Accommodation of pressures for residential (including affordable housing) and business development.

5. Achievement of the various development objectives listed in this paper, including energy and resource conservation, minimization of motor traffic, public access to at least selected areas for recreational purposes, construction of walking and bicycle paths to the nearby Appalachian Trail and the New River (connecting urban Blacksburg through the rural Prices Fork area to the primeval Jefferson National Forest and New River areas), etc.

**Planning Roles of Blacksburg, Montgomery County, the New River Valley Planning District Commission (NRVPDC), the Commonwealth of Virginia, and the U.S. Government.** Town and County planning authorities must collaborate to achieve these objectives, since only a portion of the Price's Fork corridor lies within Blacksburg boundaries. It would be logical to think of the continuance of the planning corridor all the way to the banks of the New River, a residential and recreational area that lies at the boundary of Montgomery and Pulaski Counties. Blacksburg's plan for low-density development in the area should be reconsidered in the light of implications for elimination of working farms and requirements for transportation and utilities, and as seen in relation to Yaro's ideas about cluster development. The existing townhouse clusters of Oak Manor and Hethwood provide good working models.

The NRVPDC (Montgomery, Giles, Pulaski, and Floyd Counties, and the City of Radford) should facilitate multicounty transportation and walking path linkages as needed, and should protect against suburban/exurban encroachment on rural and primeval areas of significant agricultural, environmental, or historic importance. The Virginia Commission on Population Growth and Development should assist in developing a Commonwealth growth management structure for strategic planning and implementation of overall settlement policies and programs in coordination with regional and local authorities. The Commonwealth has a special responsibility for protection of selected areas such as the New River and the natural environment surrounding Virginia Polytechnic Institute and State University.

The federal responsibility cannot be overlooked. Its orientation also should be to the larger implications of settlement location and density, with special attention to preservation and conservation of selected historic and natural areas of social significance.

## RESEARCH NEEDS FOR IMPROVED SETTLEMENT POLICY IN VIRGINIA

1. Statewide inventory of urban, rural, and primeval landscapes for conservation and preservation.

2. Analysis of energy requirements of existing and proposed typical settlement patterns in high- and low-density urban and rural landscapes, with a view to energy savings by appropriate siting, use of planted windbreak barriers, and dwelling design parameters. This analysis should also include an inventory of current energy costs of transportation and alternatives, assessment of alternative community/regional energy production, and distribution models.

3. Design and testing of experimental urban "life support" communities that are energy efficient, promote ecological diversity and social interaction, reduce air and water pollution, promote low maintenance costs, and are consistent with regional environmental and social imperatives. Projects might include design approaches to "natural" air-conditioning, urban agriculture, community gardening, water conservation and management, and air quality in streets with and without trees.

4. Resolution of problems associated with local land-use regulatory approaches that tend to relegate issues such as necessary affordable housing, waste disposal, environmentally objectionable industrial processes, objectionable traffic noise and pollution, welfare support associated with low income populations, and educational needs associated with populations in child-rearing years to adjoining or other local jurisdictions. The "not-in-my-backyard syndrome" is inconsistent with the ideal of regional connectedness.

## LITERATURE CITED

Brown, L. R. 1990. State of the world. Norton, New York.

Commonwealth of Virginia. 1989. Virginia outdoors plan. Department of Conservation and Recreation, Richmond.

Lewis, C. A. 1979a. Healing the urban environment: A person/plant viewpoint. APA Journal 330–338.

Lewis, C. A. 1979b. Plants and people in the inner city. Planning, Mar.:10–14

MacKaye, B. 1928. The new exploration: A philosophy of regional planning. Harcourt Brace. (Reprinted in 1970 by Univ. of Illinois Press.)

Price's Fork Road Corridor Study Committee. 1985. Prices Fork Road Corridor Study. Blacksburg, Virginia.

Virginia Growth Management Forum. 1989. Population growth and development: Meeting the challenge of the eighties. Alexandria, Virginia.

Yaro, R. D. 1988. Dealing with change in the Connecticut river valley: A design manual for conservation and development. Lincoln Institute of Land Policy and Environmental Law Foundation, Cambridge, Mass.

CHAPTER 27 – ABSTRACT

# Celebrating City Gardeners

Bilge Friedlaender

Assistant Professor, Design of the Environment, University of Pennsylvania

## ABSTRACT

> The sun and moon cling to heaven, and grain, grass and trees cling to the earth.
>
> The *I Ching*

*Li/The Clinging Fire*
Fire clings to the wood it burns.
Its light and heat is the burning of the wood
The Garden is the wood of the gardener
The Gardener's light comes from making the garden.
Trust is the wood of the creative process,
Its flowering depends on the trust that grows between participants. From a place of openness, one can go to the unknown places of one's being to bring forth a making that is alive.

> The *I Ching*

City gardeners are able to see the earth through piles of rubble and concrete. With their hands, they liberate the buried earth to create gardens that bear food for the body, flowers for the spirit.

Philadelphia city gardeners share their vision of the possible garden in the impossible environment of alienated city blocks. They open their hearts and hands to work the earth, clinging to it as fire clings to wood.

The seed on the wind does not discriminate where it falls. It finds a home even in the crack of the concrete. This is what a city gardener knows.

SECTION V

# EXPLORING A SPECIFIC APPLICATION: HORTICULTURAL THERAPY

CHAPTER 28

# Prescribing Health Benefits Through Horticultural Activities

Richard H. Mattson, HTM

Professor, Horticultural Therapy, Kansas State University

## INTRODUCTION

As I begin this discussion on specific applications of horticulture to special population groups, I am reminded of the story concerning a 5-year-old girl who was writing a short letter to her friend. She stopped and asked her mother, "How do you spell the word, 'whale'?" Her mother spelled out the letters, "W...H...A...L...E," then began wondering why her child would be using this particular word. She asked her daughter, "Why did you want to use the word 'whale'?" The child responded, "I'm writing a letter to my friend, telling her to get whale." Obviously, she was from southern Kansas, but she was concerned about her friend's health and she was taking action.

Just like the message in this child's letter, we also need to be concerned with issues relating to wellness and improving the quality of life. We need to begin the process of letting people know about one of the "best kept secrets" around . . . gardening is one of the most healthful activities known.

Professional horticultural therapists are capable of helping special people adapt, cope, develop, and expand their personal abilities and potentials. We need to begin the process of spreading the news that horticulture is good for you. Students at Kansas State University are using the phrase, "If it's horticulture, it's therapy." A retired professor in the Department of Horticulture used to say, "Eventually, horticulture is something that grows on you." He was, of course, a turfgrass specialist.

## BACKGROUND AND DEFINITIONS

Dr. C. F. Menninger in the early 1900s, and later his sons, Dr. Karl and Dr. Will Menninger, at the Menninger Foundation, Topeka, Kansas, used the term "horticultural therapy" to refer to gardening activities conducted with psychiatric patients. Rhea McCandliss, the first horticultural therapist at the Menninger Foundation, wrote the following comments about "What is Horticultural Therapy?" in 1975:

> I have never really found a nice, concise answer of what is Horticultural Therapy, but find myself launched into a description of what it can do for such a variety of patients, each with a different problem and each being guided to some resolution of his problem by his experience in horticultural therapy.
>
> I once asked my good friend and mentor, Dr. Karl Menninger, what his definition of horticultural therapy would be. He cupped his hands up on each side of his face and said as he lowered them, "It takes the blinders off of patients and gives them a wider horizon."
>
> I believe that is just what horticultural therapists do, they widen the awareness of patients to the natural world. They open the eyes of those who have been going through the world without seeing what they are looking at. They bring to people who have never known the joy and gratification of growing plants the basic feelings of satisfaction and thus increase and enhance their self-image.

It was through a cooperative agreement with the Menninger Foundation that Kansas State University became involved in providing a curriculum for training horticultural therapists in 1971. Today, both the B. S. and M.S. degrees are offered in this specialization at Kansas State University. Dr. Karl Menninger made the following comments concerning this program:

> I believe strongly in this program of training in Horticultural Therapy. It brings the individual close to the soil, close to Mother Nature, close to beauty, close to the mystery of growth and development. It is one of the simple ways to make a cooperative deal with nature for a prompt reward.

In 1982, Kansas State University staff members working on a National Institute of Mental Health grant, wrote the following synthesis of several definitions for the term "horticultural therapy":

> Horticultural Therapy may be simply defined as the sharing of the experience of plants between the therapist and the patient/client. Their interaction creates an environment which is beneficial in mediating patient/client dysfunction because plants are universal to the human experience and symbolize a synchronization of human life with that of the Earth.

## TRAINING PROFESSIONALS

A special faculty committee at Kansas State University has been developing a "core" curriculum for all students. The current proposal includes approximately two years of liberal arts, natural sciences, and behavioral science courses—all strikingly similar to course requirements of the horticultural therapy program. If approved, the "core curriculum" plus two years of agriculture, horticulture, and horticultural therapy course work would round out the horticultural therapy curriculum. A part of this four-year curriculum is a six-month internship completed in the senior year. In addition to the B.S. degree program, approximately 20 percent of our current enrollment is in the M.S. degree program.

Originally, the horticultural therapy program trained students to work in the mental health field and the principal training, internship, and placement of graduates were within psychiatric hospitals. This is not the case today. Six specializations are offered in this field, including community-based programs, gerontology, corrections, developmental disabilities, special education, and mental health. The term "horticultural therapy" does not apply to each of these specializations. More appropriate career titles may include, "horticultural instructor," "community horticulturist," "activity therapist," or "vocational rehabilitation specialist." Horticulture provides educational, social, and vocational needs, as well as psychological benefits.

Horticultural therapy short courses are held annually by Kansas State University to provide current information to professionals, students, and volunteers. In recent years, topics have included holistic health, creative horticulture, and community-based horticulture. This annual event has drawn hundreds of participants, and proceedings have been published of each event.

## THE DYNAMICS OF HORTICULTURAL THERAPY

Two major papers were published during 1978–1979 regarding the benefits and prescriptive use of horticulture for special population groups. The first of these papers on the dynamics of horticultural therapy was written at the Menninger Foundation; the second was prepared from observations made of psychiatric patients at Vermont State Hospital.

At the Menninger Foundation, Dr. Ira Stamm and Andrew Barber (1978) outlined the dynamics of horticultural therapy in terms of specific ego functions and experiences occurring during the process of gardening. These include the following ten items:

1. Planning: This involves deciding what should be grown, the timing and scheduling of the plant, and the sequencing of the procedure.

2. Preparation: Working the soil, getting tools together, and starting the process requires involvement and anticipation of things that will happen.

3. Measuring: Linear relationships, volume measurement, and counting skills are all required to plant a garden or to transplant a seedling. These activities involve use of the analytical left hemisphere of the brain.

4. Regularity: Nurturance and continual care are required on a regular basis in order for plants to survive.

5. Creativity: The artistic aspects of horticulture allow clients to be individualistic, self-expressive, and aware of the universal language of flowers and plants.

6. Impulse management: Delaying gratification is a skill to be learned in the garden. It takes time for a plant to germinate, grow, develop, flower, and produce harvestable fruit.

7. Anxiety and tension reduction: Through physical labor, a client may be able to reduce tension and fear.

8. Response to success or failure: If the harvest is good, does the patient consume the product, sell it for profit, or save it for a rainy day? If the harvest is a failure, does the patient externalize the blame onto the weather, the therapist, or his/her peers?

9. Frustration tolerance: Learning to live with the unexpected is another lesson taught in the garden. Invasions of insects, adverse weather, and other uncontrollable events can frustrate a gardener.

10. Response to basic instincts: "Foremost, horticulture touches on life and death and the broad repertoire of human emotions that life and death stir in all of us. In between the beginning and end of life is a phase of growth and decline, of

developing new capacities and skills and losing them with age. The issue of nurturing life, or caring for it and ministering to it in health and in illness, through good times and bad, are emotions and feelings experienced by all of us. Again, the experience the patient has with his plant world, can be a microcosm for the experiences in every day life" (Stamm & Barber, 1978).

## MEETING HUMAN NEEDS THROUGH HORTICULTURAL THERAPY

Dorothea Slayton (1978–1979) described 15 ways that horticulture was used by the Activity Therapy Department at Vermont State Hospital to provide benefits to clients. In abbreviated form, they are as follows:

1. Activities of daily living: By establishing daily routines in the garden, clients better understand the importance of cleanliness and health in their own lives.

2. Ability to focus on task: Through sensory stimulation, the client is motivated to become involved in the gardening activity.

3. Enjoyment: By correct selection and adaptation of activities, the horticultural therapy insures that the client has an enjoyable experience.

4. Work and frustration tolerance: Begin with simpler gardening tasks and gradually build to higher levels of job skills. Task analysis procedures may be helpful with challenging work.

5. Socialization, cooperation: Sharing gardening experiences and plants provides meaningful interaction.

6. Behavioral control: Aggressive behaviors can be constructively channelled through physical work. Work with nonthreatening plant materials in a comfortable environment encourages relaxation.

7. Leisure time skills: Plant care and related activities can effectively occupy time.

8. Reality orientation: Understanding the relationships of seasonal changes and plant growth and development are fundamental human perceptions.

9. Self-esteem: Short tasks with immediate, obvious results help people with low self-esteem.

10. Self-expression: Floral or landscape design helps people to make decisions and use plants in a creative way.

11. Fine/gross motor skills: Many examples could be given of how we can use horticulture to develop these skills. Eye-hand coordination, physical conditioning, and work endurance can be evaluated with various horticultural activities.

12. Independence: Clients should be encouraged to make independent decisions and to accept the outcome of their actions.

13. Assertiveness: Through experience with a variety of horticultural activities, clients gain experience and an ability to become more involved with decision making and eventual outcome.

14. Intellectual stimulation: Curiosity may be aroused by sensory stimulation during the horticultural activity. Investigation of the natural world occurs during the process of gardening.

15. Values clarification: By developing a trusting relationship with a client, the therapist is able to provide a setting in which problems and issues can be addressed.

## *BEGIN BY UNDERSTANDING ONESELF*

Working in an academic background to train future horticultural therapists, I have found it essential to have students focus on personal attitudes, beliefs, and experiences. It is necessary to develop a sense of empathy, not sympathy, in working with people with disabling conditions. To develop this empathy, we must examine what we know and have been told about the disabled person, and experience the disabling condition through simulation. That is why we have horticultural therapy students in wheelchairs on campus; why we are designing and building an accessible landscape garden; why our student chapter of the American Horticultural Therapy Association has established an enclave with a local developmental disabilities workshop and is now employing eight mentally and physically disabled adults. Plant sales are funding the entire project, and students are involved in developing vocational training activities for the developmentally disabled clients. To overcome barriers of communication and environment, we must confront our fears and inadequacies directly, then use our creative abilities to improve our professional potentials.

## *PRESCRIPTIVE HORTICULTURE*

One exercise I have used in an introductory horticultural therapy class involves an evaluation of personal perceptions and uses of horticulture. The following form is used as part of this exercise:

### Introductory Horticultural Therapy—Student Exercise

Based on your personal experiences and value judgments, indicate a horticultural and nonhorticultural activity or object you would use to alleviate the following situations:

| Situations | Horticultural | Nonhorticultural |
|---|---|---|
| 1. You need to do light aerobic exercise. | | |
| 2. You are hungry and need a light snack. | | |
| 3. You are catching a cold. | | |
| 4. You need to relax. | | |
| 5. You are feeling depressed. | | |
| 6. You are very angry about something. | | |
| 7. You buy a gift for a best friend. | | |
| 8. You express sympathy after the death of a close friend. | | |
| 9. You want to give an elderly friend something special. | | |
| 10. You buy yourself a special gift. | | |

For each of the ten situations, please underline the activity that you would do. If each item under the horticulture column is given one point, any score of eight or better indicates that you know how to use horticulture to promote your health.

## APPLICATION OF PRESCRIPTIVE HORTICULTURE

On the above questionnaire, each question addresses a basic human need. Individual responses will differ. Let us look at each question and consider specific applications.

1. Aerobic exercise: Many forms of gardening exercise are available, ranging from fine motor to gross motor activities. For the physically disabled and elderly, fine motor projects may include pot washing, mixing soil, transplanting, or floral design. For people with more physical abilities, gross motor tasks may include pulling weeds, pushing a rototiller, raking leaves, or making a compost pile. Endorphin highs can occur in the garden, similar to those produced by jogging or bicycling. The majority of people are unaware of the aerobic, non-competitive nature of gardening. Clients may indicate that they would prefer to do light warm-up exercises or sports-related aerobic exercise. Gardening is still considered to be work by many people, whereas recreation is a more acceptable form of exercise. Gardening burns off excess calories and improves physical conditioning.

2. Dieting and weight control: Medical and nutritional studies report that the products of the vegetable and fruit garden are health-promoting foods. Rich in vitamins and minerals, high in fiber, low in calories and cholesterol, the carrots, green and yellow vegetables, and fruits grown in our gardens produce healthy bodies. A clove of garlic a day will prevent arteriosclerosis; cranberry juice helps clear up kidney infection; broccoli and cabbage prevent cancer of the colon. The list goes on and on. Even with medical evidence of the health-promoting effects of certain foods, the most popular light snacks are candy bars, pop, pie, and cookies. Apples, peanuts, fruit juices, and dried fruit may be purchased in some vending machines. Will we ever see carrots sold in a vending machine?

Food is a natural reinforcer and appeals to the need of most clients for sensory stimulation. Taste, touch, smell, sight, and sound can be incorporated into many food-related activities. Cooking classes, food preservation, tasting new fruits and vegetables, and garbage gardening projects (i.e., growing plants from the left-over parts, such as carrot or pineapple tops or grapefruit seed) appeal to many special client groups.

3. Health maintenance: Did you wake up this morning with a sore throat and wonder whether you were catching a cold? If so, what action did you take? If you prescribed a horticultural solution to this problem, you may have eaten an orange or grapefruit. Many believe that by maintaining a recommended daily allowance of vitamin C, getting plenty of sleep, and watching your nutrition, your body is able to ward off cold and flu viruses. I grow and can my own vegetable juice made from tomatoes, green and hot peppers, celery, and onions with a dash of tabasco sauce. I rarely get a cold with this megadose of vitamin C. The common alternatives are to take expensive antibiotics and antihistamines, miss work, and feel miserable for a few days.

4. Stress reduction: Doxon, et al. (1987) reported that clients working in a greenhouse training program had lower stress than clients who worked at an adult training center. These studies measured physiological stress indicators, including blood pressure, heart rate, and electrodermal (sweating) responses. Interior environments appear to be more relaxing and comfortable spaces with the presence of green or flowering plants. Stress reduction may be an active aerobic process or a passive meditative activity. Rather than participate in a horticultural activity, the majority of people would rather listen to music, watch television, or attend an athletic event. Our society lacks sufficient green space, city parks, community gardens, and designed interior spaces for plants. Shopping malls are popular spaces because many modern designers have included green space within the structures.

5, 6. Appropriate emotional responses: For many, an appropriate means of releasing emotions such as depression and anger is difficult to find. Appropriate horticultural prescriptions for depression may include making a floral arrangement, nurturing garden plants, sharing plant produce with a friend, or reflecting on events happening within the garden.

Anger can be sublimated through gardening activities. Physical pursuits such as chopping weeds or cutting fireplace logs with a handsaw produce needed end results and allow the angry gardener to release pent-up frustrations. Constructive and destructive activities are part of the gardening cycle.

7, 8. Communicating with plants: The universal language of flowers and plants allows gardeners to express their emotional feelings in a social context. Feelings such as love can be easily expressed and understood through the use of flowers. Red roses have precisely the same meaning in Manhattan, Kansas as they do in Manhattan, New York, or in any other location on earth. Their expression of love is clearly understood.

Grief is universally expressed through the use of flowers. This ancient association of flowers and death dates to 60,000 years ago when Neanderthal men who lived in caves in northern Iraq used flowers in burial ceremonies. Many of the plants, including the yarrow, cornflower, grape hyacinth, and hollyhocks, were harvested from plants that also have herbal and medicinal properties.

9. Social interaction: One of the essential characteristics of horticultural activity is that it produces a valuable product. The worth of this product is established when it is sold or given away. For many nursing home residents or medical patients, a common remark is, "Why should I plant a garden, I have no use for the tomatoes, the apples, or the flowers." In these situations, the best response may simply be to say, "Have you considered giving this to your friend, a nurse, a volunteer aide, a family member?" If the gift is given and is accepted with a smile, horticulture assumes a social significance that cannot be undervalued. Creating, sharing, and exchanging products of the garden promote social bonding in a realistic and meaningful way.

10. Gardening builds self-esteem: To people actively involved in gardening, the process and the product of the experience have significant psychological value. It is essential that the horticultural therapist correctly direct the process of gardening to ensure success. Success is measured by the end product. Without an ample harvest, a beautiful floral arrangement, or an accessible landscape, the client's self-esteem will not grow to its full potential. The product of the garden is the natural reinforcement of the activity. The value of a bushel of tomatoes is more than just dollars and cents.

## PRESCRIPTIVE HORTICULTURE FOR FUTURE HEALTH AND WELLNESS

The decade of the 1990s holds promise that people may begin to improve the quality of life on the planet Earth. Today we live in concrete jungles with air so polluted that it is unfit to breath, we drink water poisoned with carcinogens, and we consume junk food of minimal nutritional value. We are aware of the greenhouse effect, global warming, the destruction of the ozone layer, and the long-range influence of the destruction of the rain forests and other natural areas. Clean air and water, biological pest control, soil conservation, and appropriate natural resource management seem to be challenges we can meet.

On a patient-by-patient, case-by-case basis, we as horticultural therapists have the ability and knowledge to begin improving the quality of human life and environmental conditions. Horticultural therapists have historically been organic gardeners. We recycle newspapers and aluminum cans. We are self-reliant, growing and storing our own fruits and vegetables. We need to expand our involvement in environmental education, vocational training of job skills, and issues concerning human nutrition, public health, economic development, and the environment.

Through personal experiences and by working with special groups, many of us have witnessed the remarkable processes that occur when people and communities change. As horticultural therapists we are "agents of change." We mediate the process through which people recover, adjust, or cope with impairments, disabilities, and handicapping conditions.

We must assume responsibility to inform more people about the unique prescriptive benefits of horticultural activities. It is time to take action. Horticultural therapeutically speaking, it is time to get growing.

## LITERATURE CITED

Doxon, L. E., R. H. Mattson, and A. P. Jurish. 1987. Human stress reduction through horticultural vocational training. HortScience 22(4):655–656.

Slayton, D. 1978–1979. Meeting individual needs through horticultural therapy. National Council for Therapy and Rehabilitation through Horticulture Newsletter 5(12):3,6; (1):2,6; (2):2,6; (3):3.

Stamm, I. and A. L. Barber. 1978. The nature of change in horticultural therapy. Directions '78. Proceedings of the 6th Annual Conference, National Council for Therapy and Rehabilitation through Horticulture, pp. 11–16.

CHAPTER 29

# Measuring the Effectiveness of Horticultural Therapy at a Veterans Administration Medical Center: Experimental Design Issues

James A. Azar

Counseling Psychologist, Veterans Administration Medical Center,
Northampton, Massachusetts

Thomas Conroy

Doctoral Candidate, Communications Department, University of Massachusetts

A call was made at the 1989 American Horticultural Therapy Association's National Conference to unite therapists and researchers in an effort to measure the effects of horticultural therapy. This paper describes our research efforts, the inherent difficulties in conducting an experimental study within a hospital setting, and our plans for the future.

The foundation of a "true" experimental design is the establishment of a control group that serves as a comparison measure for the treatment of the condition that is under investigation. This model requires that subjects are randomly placed in either the control or treatment group and that the decision is determined solely by chance. A major goal of randomization is to form essentially equivalent groups, i.e., no differences with respect to age, education, socioeconomic status, psychiatric history, psychopathology. The primary goal in the establishment of a control group is that it provides the researcher increased confidence in stating that the hypothesis under investigation has not been disconfirmed by such sources of internal invalidity as history, maturation, testing, regression, and selection (Campbell and Stanley, 1963).

Conducting a true experimental study in a hospital setting presents a number of difficulties, the most crucial of which is the process of randomly placing patients into a control group. In a psychiatric hospital, patients are often psychotic and/or a danger to themselves or others at the time of admission. A period of stabilization and/or observation is often needed. As a result, placing patients into a horticultural therapy program at the time of admission would not be a sound clinical decision. In addition, with the present financial stressors pressing upon the mental health system in the form of decreased funding and the use of predetermined lengths of stay through the use of Diagnostic Related Groups (DRGs), patients in an acute crisis unit are often discharged to the community or to an intermediate psychiatric unit within 30 days. For these reasons, the establishment of a control group at the point of admission is not possible.

Problems also exist in establishing a control group at the intermediate psychiatric unit level. A major difficulty with a psychiatric population is recidivism. Many of the patients that are hospitalized at a V.A. Medical Center or any public psychiatric facility have had several prior admissions. Since our horticultural therapy program has been in existence for over 10 years, the likelihood is that some of the newly admitted patients have been in the program in the past. If patients with prior experience are placed in either the treatment or control group, then the "purity" or separateness that is assumed between the groups is compromised. An ethical problem also emerges in establishing a control group in a governmental public sector facility. For example, veterans may request horticultural therapy as a component of their treatment program if in the past it had been helpful, or if they have an interest in gardening. Ethical and clinical concerns are raised if they are denied access to the program on the grounds that it disrupts the random assignment condition of the experimental design.

One possible approach to compensate for the lack of a control group is to match patients on a variety of variables. Some of the variables that would need to be considered are age, sex, education, psychiatric history, diagnosis, and alternate forms of treatment such as medication, psychotherapy, and education groups. As one can imagine, matching patients on these dimensions would be a time-consuming and exhausting process and would certainly restrict the number of patients involved in a study. Assuming that the time and resources were available, all of the relevant variables could not be confidently accounted for and kept constant through a study.

Another difficulty in measuring the effectiveness of a horticultural therapy program is the development of psychometrically sound outcome measures. During our initial review of the literature, we became aware of the need for an instrument that would be sensitive to changes. This awareness led to our attempt to develop such an instrument. After several studies, a 28-item scale with promising reliability and validity emerged (Azar and Conroy, 1989).

Due to the difficulties in establishing a control group in a hospital setting, we have temporarily suspended our desire to conduct a true experimental study. The goal of our present project is to refine our outcome measures in order to quantify the changes that we feel occur with a psychiatric population. Our hypothesis is that patients experience improvements in such areas as general psychiatric functioning, self-esteem, socialization, and the ability to work with co-workers and supervisors as a function of their involvement in horticultural therapy. We still need to determine, however, whether we have the proper outcome measures to detect changes and if those changes are perceived differently by the various members in the therapeutic process (e.g., patient, horticultural therapy staff, hospital psychologist). We would have little reason to encounter the obstacles involved in the establishment of a control group unless we had faith in the dependent variables that would be utilized.

We believe that a longitudinal study with observations from several different agents in the therapeutic process would be a fruitful next step. The patients would fill out a self-esteem scale at the time they begin the program, and the staff would complete the 28-item Horticultural Therapy Questionnaire that we designed, mentioned above. A psychologist outside of the program would provide a general assessment of psychiatric functioning via the Global Assessment Scale (Endicott, et al., 1976). These measures would be repeated at several times (e.g., 30 days, 60 days, and discharge). A discharge interview would also be conducted to obtain the patient's perceptions of the positive and negative contributions of horticultural therapy in their overall treatment plan. We would hope to develop more appropriate future outcome measures from these exit interviews.

Several beneficial outcomes may be possible from this type of study. It would be the first known attempt in the horticultural therapy field to use multiple-dependent or outcome measures that incorporate the patient, a horticultural therapy staff, and a hospital psychologist. The study may provide evidence that change is perceived at different rates by the various agents in the treatment process. If that is the case, it would provide researchers with a better understanding in the future as to what outcome measures need to be included in a study to detect the changes that may be occurring. The data could also be analyzed with respect to diagnosis, since there has been speculation in the literature that certain symptoms respond to different aspects of horticultural therapy (Spelfogel and Modrzakowski, 1980). This study may facilitate subsequent research with regard to that question if it is found that the rate of change is contingent upon the diagnostic category.

Nevertheless, there are clear limitations with this type of research design, the most important of which is that if positive changes are observed over time, those changes cannot be solely attributed to horticultural therapy in a scientific sense. Without a control group, the alternative hypothesis that these changes are the result of some other factor or combination of factors is not ruled out. For example, it can be easily argued that improvements in self-concept, work performance, socialization, or psychiatric functioning as evidenced on the outcome measures may be the result of medication, psychotherapy, the structure and nurturance of a hospital setting, or the cyclical nature of a psychiatric illness.

The question may remain for some as to the rationale in performing the proposed study if the results do not allow us to rule out conflicting explanations for the changes that may occur. Our rationale is that the study would provide an intermediate step that helps us better understand and measure the process of change before a more formal design is arranged. We feel that it continues the task of the 1990s set forth by the horticultural therapy community, which is to measure and better understand the effects of horticultural therapy in various populations.

## LITERATURE CITED

Azar, J. A. and T. Conroy. 1989. The development of an empirical instrument designed to measure the effects of horticultural therapy. Journal of Therapeutic Horticulture 4:21–28.

Campbell, D. T. and J. C. Stanley. 1963. Experimental and quasi-experimental designs for research. Rand McNally, Chicago.

Endicott, J., R. L. Spitzer, J. L. Fleiss, and J. Cohen. 1976. The global assessment scale: A procedure for measuring overall severity of psychiatric disturbance. Archives of General Psychiatry 33:766–771.

Spelfogel, B. and M. Modrzakowski. 1980. Curative factors of horticultural therapy in a hospital setting. Hospital and Community Psychiatry 31:572–573.

CHAPTER 30

# Developing a New Computer-accessed Data Base for Horticultural Therapy Research at Rusk Institute

Nancy K. Chambers, HTR and Patrick Neal Williams, M.S., HTR

Horticultural Program, Enid A. Haupt Glass Garden, Howard A. Rusk Institute of Rehabilitation Medicine, New York University Medical Center

The Howard A. Rusk Institute of Rehabilitation Medicine was established in 1948 as part of New York University Medical Center. It was the first facility in the world devoted solely to the rehabilitation of individuals with physical disabilities, either congenital or acquired through illness or injury. The patients stay on average two months, but this can be extended depending on the individual's disability. Patients enter the Institute after they are medically stable from conditions such as stroke, head trauma, brain surgery, hip surgery, spinal cord injury, lung disorders, and amputations.

At Rusk, patients undergo physical and occupational therapies, speech therapy, therapeutic recreation, vocational training, and psychological support. They also receive orthotic and prosthetic services as needed. The collective goal of these treatments is to help individuals obtain their maximum degree of independence—physically, emotionally, socially, and vocationally.

Horticultural therapy has been an integral part of patient treatment at Rusk for more than 15 years. This program currently averages over 7500 patient therapy hours a year. Individuals in the treatment sessions perform all horticultural tasks necessary to maintain the Enid A. Haupt Glass Garden. They also select and propagate plants for themselves.

Goals in the horticultural therapy program are defined for each patient. They focus on functional areas, such as fine and gross motor dexterity and coordination, cognitive and perceptual skills, social interaction, problem solving, and the ability to cope psychologically and emotionally.

In spite of the program's acknowledged success in contributing to the physical rehabilitation and regained independence of patients, no research has been done on the program or its effects, because a systematic way to evaluate objectively a patient's progress over time, to compare one patient's progress with another, or to relate that progress to aspects of the horticultural therapy program alone has been unavailable. Although patient progress has always been documented, the method of recording progress has been subjective and inconsistent.

Some treatment disciplines at Rusk had developed more objective means of testing, evaluating, and recording patient progress. The lack of objectivity and consistency was more problematic in horticultural therapy, because many different patient functions are involved at any one time in horticultural activities. Whereas other disciplines might be able to focus on one particular functional area at a time, such as speech or lower extremity movement, work in the greenhouse involves many different skills at once.

New York University Medical Center recently established a new computerized Hospital Information System (HIS) to document patient progress in all areas of treatment. As part of this new system, the horticultural therapy staff designed a new recording format (Figure 1) to evaluate objectively patients' progress in relation to horticultural activities. This tool will track progress better and will indicate where treatment can be adjusted so that patients receive the maximum benefit.

It will also provide, over time, a consistent, objective data base that can be used in research on the application and benefits of horticultural therapy. Although the format was designed for use with patients who have physical disabilities, we believe it can also serve as a prototype for evaluating the effectiveness of horticultural therapy with other populations as well.

The new reporting format consists of a computer screen on which therapists enter documentation on each patient every two weeks. The various functional areas are broken down into seven broad categories: mobility, physical/perceptual abilities, writing, social interaction, cognitive ability, emotional status, and avocational interests. A total of 70 possible functions can be graded on a scale from 1 through 5. These functions are designed to correlate to various assessments of patients in other treatment programs at Rusk. Space is also provided for subjective notation summarizing current functional status and for specification of the projected treatment plan.

A rating of 5 on the scale indicates a patient's ability to perform the function independently. A rating of 1 indicates an ability to function only with a maximum amount of assistance. Ratings in between indicate varying degrees of independence and assistance needed. We are currently developing a detailed protocol that will clearly define each rating point on the scale. This should provide a high degree of consistency in ratings among different recorders and ensure a more objective evaluation. A rating of Not Applicable is given when a function does not apply to a patient or when meaningful measurement is not possible; for example, if a severe stroke makes bilateral task completion impossible or a person is unable to interact with others due to extreme aphasia or a foreign language.

Patients are evaluated on the basis of how well they perform normal horticultural activities in the greenhouse. No evaluations are administered, such as asking a patient to sort pots according to size and shape, for two reasons: First, one of the real values of work in the greenhouse is having patients apply skills they have acquired in other parts of the treatment program in practical applications—completing a task that needs to be done. Second, we felt that an ability to observe progress in a natural way, during the patient's routine activities in the greenhouse, would give us a more realistic picture of the individual's ability to function at home.

This new format for evaluating and recording patient progress can make an important contribution to future research in horticultural therapy. Over time, the accumulated data on a wide range of patients with varying physical disabilities should enable researchers better to understand, and to document, how horticultural activities relate to specific physical, perceptual, cognitive, social, and emotional functions. This, in turn, will increase our understanding and appreciation of horticultural therapy, not only as a means of rehabilitation, but as a way of enhancing overall quality of life and well-being.

**NEW YORK UNIVERSITY MEDICAL CENTER**
**The Rusk Institute of Rehabilitation Medicine**
**HORTICULTURAL THERAPY**
**GROUP ACTIVITY TREATMENT PROCEDURE**
Attendance:

Name: _____ Age: ____
Chart No.: _____ Room ____
Diagnosis: _____
Disability: _____
Physician: _____

KEY
5 — independent
4 — consistent/ w/supervision
3 — minimum assist
2 — moderate assist
1 — maximum assist
N/A — non-applicable

M ____ T ____ W ____ TH ____ F ____
  ½  1   ½  1   ½  1    ½  1    ½  1

M ____ T ____ W ____ TH ____ F ____
  ½  1   ½  1   ½  1    ½  1    ½  1

TREATMENT GOALS:

| MOBILITY | 5 | 4 | 3 | 2 | 1 | N/A |
|---|---|---|---|---|---|---|
| Comes independently to HT session | | | | | | |
| Comes on time to HT session | | | | | | |
| Able to maneuver within the greenhouse | | | | | | |

| PHYSICAL/PERCEPTUAL ABILITIES | | | | | | |
|---|---|---|---|---|---|---|
| Performs horticultural tasks: bilaterally | | | | | | |
| left-handed | | | | | | |
| right-handed | | | | | | |
| Able to grasp/release tools during activity | | | | | | |
| Able to manipulate tools | | | | | | |
| Able to manipulate plant and non-plant materials | | | | | | |
| Able to fill pot accurately | | | | | | |
| Able to center cuttings in pot | | | | | | |
| Able to place cuttings in dibble holes | | | | | | |
| Able to place plant in upright position | | | | | | |
| Able to place scissors in proper position for cutting | | | | | | |
| Selects correct stem when cutting from multi-stem plant | | | | | | |
| Able to space cuttings in whole pot | | | | | | |
| Able to work with plant materials: in front | | | | | | |
| above head | | | | | | |
| to sides | | | | | | |
| Able to water plants with: 1 lb. can filled | | | | | | |
| 2½ lb. can filled | | | | | | |
| Able to water plants accurately | | | | | | |
| Able to wash hands &/or nails in sink | | | | | | |
| Endurance permits completion of horticultural tasks | | | | | | |
| Able to work on specimen plants (over 6" pot) | | | | | | |
| Able to complete horticultural tasks correctly | | | | | | |
| Able to find all materials on table for task | | | | | | |
| Able to control physical problems/pain during task | | | | | | |

| WRITING ABILITY | | | | | | |
|---|---|---|---|---|---|---|
| Writes own name on plant label | | | | | | |
| Writes date on label | | | | | | |
| Writes plant name on label | | | | | | |
| Handwriting is legible | | | | | | |

| SOCIAL INTERACTION | 5 | 4 | 3 | 2 | 1 | N/A |
|---|---|---|---|---|---|---|
| Hearing impairment limits socialization | | | | | | |
| Foreign language limits socialization | | | | | | |
| Aphasia limits socialization | | | | | | |
| Initiates interaction w/therapist | | | | | | |
| Interacts once approached | | | | | | |
| Interacts with peers | | | | | | |
| Interacts appropriately with others | | | | | | |
| Fits easily into group | | | | | | |
| Responds accurately/appropriately to questions | | | | | | |
| Able to make self understood | | | | | | |
| Able to discuss physical conditions realistically | | | | | | |

| COGNITIVE ABILITY | | | | | | |
|---|---|---|---|---|---|---|
| Follows verbal &/or written directions: 1 step | | | | | | |
| 2 step | | | | | | |
| more | | | | | | |
| Follows demonstrated directions: 1 step | | | | | | |
| 2 step | | | | | | |
| more | | | | | | |
| Able to remember task sequence between sessions | | | | | | |
| Able to focus on task | | | | | | |
| Able to maintain attention span for 1 hour session | | | | | | |
| Able to shift from one task to another | | | | | | |
| Able to control behavior to complete tasks accurately | | | | | | |
| Follows safety precautions | | | | | | |
| Understands basic horticultural concepts | | | | | | |
| Understands purpose of HT treatment | | | | | | |
| Able to adhere to time schedule | | | | | | |
| Aware of seasons, weather, whereabouts | | | | | | |
| Able to overcome problems encountered during tasks | | | | | | |

| EMOTIONAL STATUS | | | | | | |
|---|---|---|---|---|---|---|
| Is willing to try new activities | | | | | | |
| Seeks assistance when appropriate | | | | | | |
| Has confidence in horticultural tasks attempted | | | | | | |
| Perseveres on difficult tasks | | | | | | |
| Able to control emotional status during activity | | | | | | |

| AVOCATIONAL INTERESTS | | | | | | |
|---|---|---|---|---|---|---|
| Selects own plant material for propagation | | | | | | |
| Shows interest in learning cultural practices of plants | | | | | | |
| Maintains own plants propagated during treatment | | | | | | |
| Anticipates taking plants home | | | | | | |
| Visits greenhouse on own time at least 3x/week | | | | | | |

CURRENT FUNCTIONAL STATUS:

PROJECTED TREATMENT PLAN:

_____
*Therapist*

HT./RIRM 1/90

**Figure 1.** The new computerized Horticultural Therapy Evaluation Form used at New York University Medical Center to document patient progress. (Permission to reproduce granted by the authors.)

CHAPTER 31

# Research:
# An Imperative for Horticultural
# Therapy and Third-party Payment

---

Sally Hoover, HTR

Horticultural Therapist, Urban Horticulture Department, Chicago Botanic Garden

Increasingly, the profession of horticultural therapy faces a very serious and demanding challenge: to prove to the medical community the effectiveness of horticultural activities as treatment interventions—in short, that horticultural therapy "works." The quantified documentation of this can potentially set off a series of chain reactions that will have a lasting impact on the profession itself.

The culmination of the chain reaction is third-party payments: payments from public and private insurance for therapeutic services rendered. What sets off the reaction is a body of research quantifying the effects of horticultural therapy. In order to qualify for third-party reimbursement, horticultural therapists must be able to document that their work is important and unique, to show why insurance—or anyone else—should pay for their services at a time when the medical community is under severe financial constraints.

There has been some seminal writing by Lewis (1973), Kaplan (1973), Relf (1981), Stamm and Barber (1978), and others on the theories underlying horticultural therapy, but the body of research measuring actual change is small. Some useful studies have already been done. Researchers have developed an instrument to measure cognitive, emotional, social, and psychological effects of a horticultural therapy program (Azar and Conroy, 1989). The Horticultural Hand Capacities Test, developed by Gallagher and Mattson (1986), correlates horticultural tasks with similar tasks on the Physical Capacities Evaluation of Hand Skill Test used in occupational therapy. Bunn's study (1986) has laid groundwork for assessing the role of plants in social bonding. All these point the way toward useful assessment tools for horticultural therapists, but we now need to take these tools and others that may be developed, establish their reliability and validity by repeated testing, and use them in scientific research.

Research in allied adjunctive therapies, for instance, could be applied to horticultural therapy. Cofrancesco (1985) measured the effect of music therapy on hand-grasp strength and functional task performance in stroke patients. Her initial research found that the subjects "improved considerably," with enhanced bilateral movement and functional and coordination skills. Such a study might be further developed by using horticultural therapy as the intervention.

Questionable projects might be redone. For example, a study on the effect of a horticultural therapy program on children with cerebral palsy was conducted by Ackley and Cole (1987) at a site where there was a part-time, volunteer horticultural therapist who did not "enjoy privileges in communication with other professional staff." The results, as one might guess, were not supportive of horticultural therapy. Because the study had a small group size and was conducted at just one facility, however, the researchers—fortunately for the profession—recommended further study rather than abandonment of horticultural therapy programs.

New research that could be considered includes the following: (1) in the cognitive domain, measurable changes in carryover or reality orientation; (2) in the physical domain, measurable changes in upper extremity strength and movement or fine motor movements; (3) in the psychological domain, increased display of appropriate behavior or acceptance of responsibility. These are a few ideas; the possibilities are limitless.

By expanding on the image of the chain reaction, we can postulate that tested, verifiable results could provide the basis for the national organization, the American Horticultural Therapy Association (AHTA), to develop professional standards and protocols of practice and concomitant certification. Schools and universities would train students based on these standards, which the students could then apply in the workplace (Aguilar, 1989). Zandstra (1987) has emphasized the importance of evaluations in securing funding and in determining the effectiveness of a horticultural therapy program. Indeed, sound documentation can lead to an increase in the prescription of horticultural therapy for treatment, and once horticultural therapists are able to bill separately, will give them greater autonomy, professional integrity and strength, and more control in the clinical setting.

Payment for services ensures the continuation of (and potential increase in) programs, thus resulting in increased jobs. The demand for more jobs implies the need for more qualified practitioners, which in turn will increase the demand for training. This may well alert schools and universities to the possibility that horticultural therapy could be a vital curriculum addition, thereby providing a recruiting tool for horticultural programs that seem to be in trouble (Childers, 1989). The introduction of new markets for horticultural products and services also benefits industry, from interior plantscapers, to growers, to breeders looking for better plants.

We at the Chicago Botanic Garden would like to participate in a collaborative research effort. We work in a number of health-care facilities, training staff in horticultural therapy programming, and thus we can provide both the horticultural therapists and the populations to be studied. In fact, we already have one facility, part of an esteemed research/teaching complex, willing to serve as a research site for working with patients in their geriatric rehabilitation program. We are actively seeking researchers and sources of funding.

Obviously, the road to third-party payments is a long one, but it seems clear that any steps taken—and the more, the better—will result in gains for the profession, the horticultural industry, and most importantly, the people whose lives horticultural therapists affect.

## LITERATURE CITED

Ackley, D. and L. Cole. 1987. The effect of a horticultural therapy program on children with cerebral palsy. Journal of Rehabilitation 53:70–73.

Aguilar, T. E. 1989. Collaboration: A simplification. NTRS Report. Handout, IL Park and Recreation

Assoc./IL Therapeutic Recreation Section, 1989 Annual Meeting.

Azar, J. A. and T. Conroy. 1989. The development of an empirical instrument designed to measure the effects of horticultural therapy. Journal of Therapeutic Horticulture 4(1):21–28.

Bunn, D. E. 1986. Group cohesiveness is enhanced as children engage in plant-stimulated discovery activities. Journal of Therapeutic Horticulture 1(1):37–43.

Childers, N. F. 1989. Accumulated philosophy on student recruiting, teaching, research, and funding in horticulture. HortScience 24(6):895–896.

Cofrancesco, E. M. 1985. The effect of music therapy on hand-grasp strength and functional task performance in stroke patients. Journal of Music Therapy 22(3):129–145.

Gallagher, M. J. and R. H. Mattson. 1986. Evaluation of arthritis using the Horticulture Hand Capacities test. Journal of Therapeutic Horticulture 1(1):31–36.

Kaplan, R. 1973. Some psychological benefits of gardening. Environment and Behavior 5(2):145–161.

Lewis, C. A. 1973. People-plant interaction: A new horticultural perspective. Amer. Hort. 52:18–25.

Relf, D. 1981. Dynamics of horticultural therapy. Rehabilitation Literature 42(5–6):147–150.

Stamm, I. and A. Barber. 1978. The nature of change in horticultural therapy. Paper presented at the 6th Annual Conference of the National Council for Therapy and Rehabilitation through Horticulture, Topeka, KS.

Zandstra, P. J. 1987. Evaluating the effectiveness of your horticultural therapy program. Journal of Therapeutic Horticulture 2(1):23–27.

CHAPTER 32

# Horticultural Therapy: Potentials for Master Gardeners

Joel S. Flagler, MFS, HTR

Agricultural Agent, Rutgers University Cooperative Extension

## INTRODUCTION

Horticultural therapists provide a service that is often categorized as nonessential. For this reason, program budgets may be far less than optimum. Nationwide, horticultural therapy facilitators must utilize alternative manpower to assist in delivering activities and projects, maintaining gardens and greenhouses, and managing overall programs.

The broad objective of the research described here has been to identify several pools of volunteers from which horticultural therapists can draw assistance. It is widely agreed that staffing can be a serious limiting factor in program operation and expansion (Norman, 1986). It is increasingly valuable for horticultural therapists to have knowledgeable, dependable assistants on a salary-free and sustained basis.

The facilities in this study include nursing homes, hospitals, mental health and rehabilitation centers, and one county jail. All are located in northeast New Jersey.

## IDENTIFYING VOLUNTEER SOURCES

In each application, volunteers play an indispensable role in the horticultural therapy program. The groups sponsoring the volunteers are diverse. They include botanical gardens, community colleges, Cooperative Extension Master Gardeners, and the Retired Senior Volunteer Program (RSVP).

### Retired Senior Volunteer Program

The services of RSVP volunteers, aged 60 and older, are available to nonprofit facilities and organizations in every state. The typical county in the northeast New Jersey study area has an average of 600 RSVP volunteers on the job in nursing homes, hospitals, and special service agencies of all kinds (Earner, 1990).

The RSVP operation is funded by the federal ACTION program, although in many applications it may be cosponsored by United Way or HUD. What is unique about RSVP is that it utilizes a portion of our population that has historically been overlooked and underestimated.

There is no limit on the time a RSVP volunteer can contribute. Many stay on the job for years, providing important continuity and long-term reliability (Earner, 1990). RSVP is operational in every state, so horticultural therapists may find volunteers widely available.

## College Cooperative Education

Community colleges often require a cooperative education (co-op) experience of their students (Rasmussen, 1989). If the students are majoring in horticulture, or another plant science, they may make effective assistants in horticultural therapy programming. With community colleges in place across the country, these co-op education students can be another widely available resource.

The cooperative education volunteers in this study are connected with a local community college and are required to contribute 180 hours. The intent is for students to gain hands-on experience in their particular professional focus.

The flexibility built into the co-op program can be a convenience for both the student and the agency providing the experience. A 180-hour obligation, for example, could be satisfied over a five-month period on a basis of two or three days per week. This valuable assistance from May through September could enable a horticultural therapy program to get through its busiest, most productive season with maximum attainment of goals and objectives.

## Botanical Gardens

Other places to look for horticultural therapy volunteers include botanical gardens and arboreta. Those offering educational certificate programs may require their students to complete an internship to gain experience in their specialty area.

At the New York Botanical Garden, students enrolled in the horticultural therapy certificate program must complete a 40-hour internship. Students work alongside registered horticultural therapists and have the opportunity to be involved with direct client contact and program management.

One deficiency of these internships concerns their duration. A 40-hour period may only allow for a superficial and limited experience. The horticultural therapist using this pool of volunteers may find a significant percentage of time spent training and orienting new interns. Further, there may be minimal opportunity for the intern to experience continuity, particularly in the area of client relationships.

## Cooperative Extension Master Gardeners

The fourth and final source of volunteers in the study group is the Master Gardener program. Master Gardeners are a specialized group of volunteers trained through the U.S.D.A. Cooperative Extension System. From its simple beginning in 1972, the program has spread to 45 states and now includes over 15,000 Master Gardeners in the U.S and Canada (Barton, 1988).

The Master Gardeners in this study group receive 100 hours of formal classroom and field laboratory training from Rutgers University professionals. Subject matter includes soil science, botany, propagation, horticultural therapy, and greenhouse gardening. In return, each student is obligated to give 100 hours of volunteer service, although many continue long after that requirement is met. There are several avenues for returning volunteer time. The table below shows the various projects in which Master Gardeners participated in 1989.

Table 1 shows the approximate number of hours, by project category, returned by Master Gardeners in Bergen County, N.J. Horticultural therapy is the third most popular volunteer project area, following the annual spring fair and the state botanical garden at Skylands.

In order to gain a better understanding of why Master Gardeners volunteer in horticultural therapy programs, we are preparing a survey. Data will be requested of Master Gardener program coordinators and volunteers in every state. The results of the survey should provide important information on the total contribution of Master Gardeners and their motives for selecting horticultural therapy over other volunteer project areas (e.g., hotline, beautification, etc.).

**Table 1.** Master gardener hours contributed in 1989 in Bergen County, NJ.

| Project area | Total no. of master gardener volunteers | Total no. of master gardener hours |
|---|---|---|
| Spring fair | 40 | 850 |
| State botanical garden | 17 | 840 |
| Horticultural therapy | 18 | 360 |
| Hotline | 22 | 300 |
| Newsletter | 6 | 90 |
| Community beautification | 20 | 60 |
| Speaker bureau | 16 | 50 |

## DISCUSSION

Potentially, all groups of volunteers may be assets to the programs they serve. Working with short-term volunteers (e.g., botanical garden interns), however, may present problems particular to certain applications. One example is in correctional facilities, where security is a priority. It is desirable to have the same individuals each time, coming on a strictly scheduled basis. Personnel changes, including volunteers, can be upsets to the participants and the administration. The need for security checks and clearances make it necessary to have as much stability and continuity as possible.

Another example of the need for long-term volunteering is seen in horticultural therapy programs serving mental health and psychiatric populations. Here, again, consistency and stability must be built into all structured activities and experiences. Changes in program personnel can be upsetting and traumatic to some individuals. Horticultural therapists in these applications must utilize volunteers who can commit to long-term, dependable service.

Master Gardeners, as a group, appear to be uniquely qualified to serve in horticultural therapy settings. They are well-trained in plant sciences and dedicated to sharing their knowledge and volunteering their time. Their long-term commitment makes Master Gardeners ideal assistants for horticultural therapy programs in diverse settings, including "sensitive" facilities such as prisons and mental health centers.

Horticultural therapy programs will continue to rely heavily on volunteers to operate. With budget restrictions a nationwide reality, it may be volunteer power that permits programs to improve and expand. Owing to their connection with land-grant colleges, Master Gardeners are a resource available to horticultural therapists in every state.

## LITERATURE CITED

Barton, B. 1988. Becoming a Master Gardener. Flower and Garden Oct/Nov:52–64.
Earner, D. 1990. Report on senior volunteers. Pers. Comm.
Norman, C. 1986. America's hardest working garden volunteers. National Gardening Nov:48–50.
Rasmussen, W. 1989. Taking the university to the people. Iowa St. Univ. Press.

CHAPTER 33

# Measuring Client Improvement in Vocational Horticultural Training

Douglas L. Airhart, HTM

Edward L. French Center, The Devereux Foundation

Kathleen M. Doutt, HTR

Bryn Mawr Rehabilitation Hospital

## INTRODUCTION

Horticultural activities incorporated into training programs can improve the behavioral and job skills of persons with a variety of disabilities (Daubert and Rothert, 1981; Hefley, 1973). These activities can be adapted in day care and sheltered workshop settings to clients with borderline competitive and competitive employment capabilities (Hudak & Mallory, 1980). The horticultural industry has experienced a shortage of trained persons to fulfill the labor requirements (Roche, 1989). Vocational horticultural training programs can provide clients trained for entry level positions in horticulture (Hefley, 1973).

Successful programs incorporate prior appraisal of a client's adaptive behavior skills, a statement of training objectives, and, if necessary, a baseline task analysis of activities. A structured work routine is followed to avoid confusion, and experienced clients assist new clients, thus building self-confidence. Clients practice skills that help them to gain a sense of job responsibility through the daily activities of plant care and greenhouse maintenance. They can improve the quality of their lives through a positive self-image and a degree of self-sufficiency (Airhart and Tristan, 1987). Few tools, however, exist for documenting the improvement and maintenance of competitive skills in horticulture.

## COMPARISON OF EVALUATION SCALES

The purpose of this paper is to address evaluative tools needed to quantify client improvement in the framework of a vocational horticultural training program. The material presented is based on the needs identified in a program designed to improve behavioral and vocational skills of persons with various disabilities. This paper is limited to the need for tools to address improvement in self-esteem, adaptive behavior, and horticultural skills. A comparison of some potentially useful existing evaluation scales is presented in Table 1.

**Table 1.** Comparison of some existing evaluation scales.

| NAME OF SCALE | Piers-Harris | Self-evaluation | COTE | Effects of HT | Melwood | HTP | Brigance |
|---|---|---|---|---|---|---|---|
| TYPE OF SCALE | Self-concept | Self-concept | Adaptive behavior | Adaptive behavior | Adaptive behavior and skill inventory | Skill inventory | Essential skills |
| CLIENT | Mixed population | Mixed population | Psychiatric | Psychiatric | Developmentally disabled | Mixed population | Mixed population |
| QUALIFIER | Oral/written 3rd–12th | Sheltered workshop capability | Day treatment | None | Entrance qualification | Vocational horticultural client | Oral/written 3rd–12th |
| MEASURES | Behavior Academic Appearance Popularity Happiness Satisfaction Anxiety | Compares client vs. therapist evaluations | Severe fluctuations in adaptive behavior | Cognitive, emotional, social, psychological | Job readiness: competitive, borderline, sheltered, day care | Improvement of horticultural skills | Curricular, vocational, personality |
| LIMITS | Not for adult clients | Minimal factors included | Psychiatric population | Not specific to horticulture | Not specific to horticulture | Skill determination | Not specific to horticulture |
| NEEDS | Modifications and testing for older clients | Similar evaluation by therapist | Modification for vocational program to measure skills | Further testing on other populations | Horticultural skills assessment | Adaptive behavior scales | Horticultural skills assessment |

### Self-Esteem

The Piers-Harris Children's Self Concept Scale (Beatty, 1969; Piers and Harris, 1969), which measures broad categories of self-esteem of children, is administered in written or oral form. Although it is not designed for adult vocational issues, its breadth makes this tool a likely candidate for modification. We have developed a client self-evaluation tool (Self-evaluation in Table 1) for vocational purposes (Doutt and Airhart, 1989). It is intended to measure discrepancies between client and therapist interpretations and to identify areas of self-esteem, but it needs expansion of categories and further testing to establish reliability and validity.

## Behavioral Adaptation

The Comprehensive Occupational Therapy Evaluation (COTE, Table 1) (Brayman et al., 1976) is widely accepted as a reliable tool for measuring severe behavioral fluctuations of psychiatric clients (Effects of HT, Table 1). The reliability of this scale, however, makes it a possible candidate for modification for vocational purposes. Azar and Conroy (1989) developed a tool for measuring the effects of horticultural therapy with psychiatric clients. The 28 items included are not specific to horticulture, but they are related to vocational abilities. This scale requires further testing with other populations to establish reliability and validity. Azar and Conroy reiterate the need for better tools for measuring client improvement.

## Skills Inventories

The Melwood Prevocational Evaluation Form (Hudak and Mallory, 1980) measures adaptive behavior and vocational skills of clients with developmental disabilities. Although it is very complete, the scale is a pre-evaluation of job readiness and is not specific to horticulture. We have developed a tool (Horticultural Training Program, HTP, Table 1) specific to horticultural skills of vocational clients (Doutt and Airhart, 1988). This evaluation needs to be expanded to include adaptive behavior scales to assess job readiness. The Brigance Inventory of Essential Skills (Brigance, 1981) is the most complete adaptable tool evaluated, capable of measuring skills of mixed populations of clients. It includes adaptive behavior and self-concept scales, and curricular and vocational skills assessment that can be selected based on the client's needs, but it does not include a section specific to horticulture.

## SUMMARY AND CONCLUSIONS

None of the tools reviewed were comprehensive in measuring improvement of self-esteem, adaptive behavior, and skill competencies in vocational horticultural training. The questions and rating scales available from certain sections of various tools could be combined, however, to develop an appropriate tool. For our particular purposes, the Melwood and HTP scale combinations would suffice. For other programs, the Brigance and HTP scales might be more appropriate.

One of the major research needs is to determine a set of goals for an evaluation scale for a vocational horticultural training. To date, none of the available literature has addressed this specific task. A set of scales made from selected combinations of existing tools could be compiled and evaluated. The need also exists for making specific advances in therapist observational skills to put these scales into operation (Zandstra, 1987). Reliable and valid vocational evaluation tools are a critical need of clients with disabilities who want to work in horticulture.

## LITERATURE CITED

Airhart, D. and J. Tristan. 1987. A horticultural training program for special students. Focus on Exceptional Children 19(5):11–12.

Azar, J. A. and T. Conroy. 1989. The development of an empirical instrument designed to measure the effects of horticultural therapy. Journal of Therapeutic Horticulture 4:21–28.

Beatty, W. H. 1969. Improving educational assessment. Association for Supervision and Curriculum Development, NEA, Washington, DC.

Brayman, S. J., T. F. Kirby, A. M. Misenheimer, and M. J. Short. 1976. Comprehensive occupational therapy evaluation. American Journal of Occupational Therapy 30(2):94–100.

Brigance, A. 1981. Brigance diagnostic inventory of essential skills. Curriculum Associates, Inc., North Billerica, MA.

Daubert, J. and E. Rothert. 1981. Horticultural therapy for the mentally retarded. Chicago Horticultural Society, Chicago.

Doutt, K. and D. Airhart. 1988. Vocational horticultural skills inventory, Levels 1–4. Unpublished.

Doutt, K. and D. Airhart. 1989. Client self-evaluation form. Unpublished.

Hefley, P. 1973. Horticulture: A therapeutic tool. Journal of Rehabilitation 3(2):20–26.

Hudak, J. and D. Mallory. 1980. The Melwood manual. Melwood Publishing, Upper Marlboro, MD.

Piers, E. and D. Harris. 1969. The Piers-Harris self-concept scale. Counselor Recordings and Tests, Nashville, TN.

Roche, J. 1989. Whither comest the people? Landscape Management 30(10):20–26.

Zandstra, P. J. 1987. Evaluating the effectiveness of your horticultural therapy program. Journal of Therapeutic Horticulture 2:23–29.

CHAPTER 34

# Horticultural Therapy in a Psychiatric Hospital: Picking the Fruit

Konrad R. Neuberger

Director, Horticultural Work Therapy Program, Langenfeld Hospital, West Germany

## INSTITUTIONAL SETTING

Langenfeld "Country Hospital" is a psychiatric state hospital for treatment of most psychiatric disorders. It lies in West Germany and is well known for some innovations. Each patient is "faced" by a medical service team, consisting of nurses, physicians, a psychologist, a social worker and a work therapist. In Germany, horticultural therapy is classified as a form of work therapy.

### Horticultural Therapy—A Prescriptive Tool

Horticultural Therapy is prescribed by the ward physician or psychologist for rehabilitative reasons (22%) or simply to help structure the patient's daily routine (78%). Twelve of our 40 wards prescribe horticultural therapy. Therefore, a regular consultation of all participants is hardly practical. As a horticultural work therapy unit, however, we have lots of time to spend with each patient—between two and six hours daily, depending on our contract. Three horticultural therapists care for up to 15 patients. Beginners receive special attention for diagnostic purposes. We produce vegetables and sell them. The work style is mainly task and group oriented.

The diagnoses and the underlying impairments of the patients are distributed as follows (based on 280 patients): schizophrenic and other nonorganic psychoses, 62%; neuroses and personality disorders, 16%; alcohol and drug abuse and connected disorders, 13%; borderline, 5%; mental retardation, 4%. Underlying the clinical descriptions are a multitude of problems or impairments, whereof matters of relation and contact are predominant.

My focus in this article is on the 10% of patients who see their stay in the hospital as a chance; they are willing to cooperate to find a solution of their problems. My idea is that the hospital can provide the time and the opportunity to look at these problems. Work in the garden, where growth is a tangible topic, provides a great chance for people with developmental deficits. There is also a chance for contact-impaired patients who work in the garden, where nothing comes off without "handling" it.

### The First Steps

Each patient who is prescribed horticultural therapy has an initial interview. I receive a survey of the patient's life, which can be relevant for the therapeutic process. We make our first contact and look for decisive situations, (which can give me a clue for treatment applications). This procedure is the same for every patient, and reveals the patient's willingness to cooperate on his problems.

### Specific Goals

The specific goals of the horticultural therapy are as follows:

1. responsibility, communication skills, better response to work obligations, improved contact, and capability to maintain it;
2. continuity, perseverance, and stamina;
3. recognition that work cannot be delayed;
4. development of new preferences and interests;
5. discover of acceptable ways to reduce stress and tension;
6. acceptable behavior;
7. maintenance of personal boundaries, one's own and other's;
8. capability of keeping contracts;
9. adaptability to work and other rhythms.

## THE GARDEN IN A FIGURATIVE SENSE—THE USE OF METAPHORS

Before I outline the application and implications of metaphors, I want to mention the methods we use.

### Our Choice of Methods

The basics of our program are as follows:

1. a preliminary interview with the patient and the physician/psychologist to share ideas, goals, and interests;
2. an agreement between the patient, the horticultural therapist, and the physician/psychologist (a contract);

3. regular meetings and the readiness of the psychologist to be available for crisis intervention.

My own education is humanistically oriented and so are some of our methods, which include Carl Rogers' "self-concept" and therapeutic attitudes; the movement-oriented Gestalt therapy with its "figure/ground," "inner dialogue," and ego-boundary concepts. The idea of a "farm-organism" and a specific approach to the human-plant relation come from the anthroposophical background of the biodynamic gardening method we use.

Our aim is similar to that of "cognitive therapy": to help change verbal and pictorial cognitions and the premises, assumptions, and attitudes that underlie the cognitions. Goals and expectations should be realistic. We do not try to change the patient's personality, values, or lifestyle.

## My Point of View

What I mostly look for is how the patients move and how they use their hands and feet. Movement tells a lot about how the patients handle other issues and how they make contact. The right task for each patient is the one with which he/she can identify. Because our patients have seldom been content or seen themselves as identities in the recent past, we reinforce them if they are whole-heartedly engaged in their work. We also need to know whether what they do matters to them. They then will tell us what comes next. They set the pace. They map out the space they want to explore. I reinforce their experimenting, and in the course of our communication, I recognize how "response-able" they are.

## Metaphors as a Bridge

Every patient has had to face changes in the past with which he/she could not cope. Connections can easily be made to gardening, where everything is constantly changing (in a less emotionally confronting way). Here plants are sown, they grow and die; each seedling needs a good contact (with the earth)—must be firmly grounded—and must be well rooted in order to grow healthy and steadily.

So we have two sides to connect, the patients' on the one side and the "plant-iful" job on the other: here is where my "thinking over" begins, in order to bridge the gap between the patient and the garden. The garden has always been a rich source for metaphors that need only to be detected and made available for therapeutic gardening. The metaphor needs to have a meaning for the patient or else it does not work.

For example, I let a certain woman in our program weed. (Weeding is a kind of sorting out, which helps the patient identify herself with the performance.) When I then asked what she was doing, she said that she was only scratching the surface. She named the movement and dropped a hint as to what extent she was involved. When I asked if that had a meaning for her life back home, she answered that she did nothing else (than scratch the surface). I told her to go on weeding with the same movement; no changes. As she was moving the soil, she could see a result. Her biography revealed that she had difficulties recognizing that what she said or did really mattered. That is why every activity in the garden is so precious. Patients can come into contact with issues they have forgotten. They themselves are able to discover parallels between their own problem and the situation in the garden. It is a perfect place to visualize and experience, to plan and to gain practice. The key to personal metaphors lies in the movement, or Gestalt, that can be developed out of a specific job at a certain moment. The patient must be open, however.

Here we have another element of therapy, the timing of the therapeutic intervention. The patient must be prepared and be ready to see the "open door," and they must be willing to give it a try. In order to explore "the field," we can bring the patients' impairments face to face with a helpful activity. By helpful, I mean activities in which the patients can find the insights themselves. In this way, the job can work as a change-inducing "eye-opener." It is extraordinarily helpful if the psychologist touches intentionally or casually the same issue (reinforcement). On the other hand, the patients can "work themselves through"; they can physically work out a problem that they have faced in a therapeutic session.

When we seek "the right work" for a patient, we have some key situations in mind. The patient will decide if he/she is ready to make use of the opportunity. I would like to illustrate this by an example. I had a therapeutic session with a very resistant woman and the psychologist. The issue was her auto-aggressive behavior (cutting herself). I had the idea that her behavior might be connected with her feelings when she felt stuck and others did not help her out. I wanted to know **how** (not why) she did the cutting at her forearm. In that way, I brought her into contact with her feelings and with what she did with her hands. She was astonished and moved. We all felt that we could not change anything by mere talking.

I went back to work with her. She had not gained a clear result from the session other than a heightened awareness of her hands. I suggested that we pick tomatoes together and hoped that the activity would yield some profit in addition to the picked fruit. She began talking to me. When she stopped (talking and harvesting), I asked her just to go on harvesting and to "concentrate on the tomatoes." She continued talking to me. She had never before seen her situation so clearly. Without further directive intervention, her ability to express herself grew and her "response-ability" rose. It was amazing how her inner development became apparent. Within seven weeks she could be discharged.

My intention was to bridge the gap between images and reality by using metaphors in Horticultural Therapy. Used carefully, they prepare "food for thought"—something that the patients deserve, too.

CHAPTER 35 – ABSTRACTS

# Horticultural Therapy and Asian Refugee Resettlement

John Tristan, HTR

Director, Durfee Conservatory, University of Massachusetts, Amherst

## Lucy Nguyen-Hong-Nhiem

Assistant Director, Bilingual Collegiate Program, University of Massachusetts, Amherst

*ABSTRACT*

The acculturation difficulties of Asian refugees were alleviated through the use of horticultural therapy activities. Vocational training in garden and greenhouse operations taught marketable skills in an environment similar to the tropical homeland of origin. Adjustment stress, fear, and culture shock became manageable and resettlement goals were accomplished.

# Evaluation of the Horticulture Therapy Program

## Sara Williams

Division of Extension and Community Relations, University of Saskatchewan, Canada

*ABSTRACT*

A horticultural therapy program in a short-term psychiatric ward was evaluated. Both social interaction and cooperative activities were fostered by participation in the horticulture group. More than 75% of the participants perceived the group to be both enjoyable and relaxing. Over half of the patients assumed responsibility for the care and watering of their plants. Most saw the program as beneficial and felt satisfaction in what they had accomplished.

# The Effect of Horticultural Therapy on the Self-Concept of County Jail Inmates

Jay Stone Rice

Community Institute for Psychotherapy, San Rafael, California

## ABSTRACT

This study will explore the impact of horticultural therapy on San Francisco County Jail inmates. An organic farm and greenhouse have been developed as part of San Francisco's "new generation" program and treatment facility.

This facility reflects a new direction in corrections that emphasizes a positive environment and direct supervision by custodial and treatment staff. The role of the horticultural therapist and the use of the natural environment as a model for growth and responsibility will be discussed. This study will address the relevance of poor self-concept and low self-esteem as important representative characteristics of county jail inmates. Research measures will explore the effect of the horticultural therapy program on the self-concept of participating inmates.

# SECTION VI

# *RESEARCH IMPLEMENTATION*

CHAPTER 36

# Conducting the Research and Putting It Into Action

Diane Relf

Associate Professor of Horticulture, Virginia Polytechnic Institute and State University

## INTRODUCTION

The implementation of a program of research to understand People-Plant Interaction[1] extends beyond the researchers to include research funding sources, communicators and educators, horticultural producers and users of the research data, and individuals who benefit ultimately from the research. Such a program of research must, by definition, be interdisciplinary in nature.

## ANALYSIS AND JUSTIFICATION

The first step in undertaking a research initiative is to collect baseline research data and determine who is currently involved in the particular type of research. This symposium documents the existence of a growing body of knowledge and brings into sharp focus the lack of involvement by horticulturists in this area of research.

The research leadership provided by Rachel and Stephen Kaplan and Roger Ulrich and their colleagues sets clear standards for the type of interdisciplinary research that could be developed between horticulturists and researchers from diverse fields. The major portion of the research conducted in environmental psychology, landscape architecture, forestry, geography, and other disciplines documents the passive response of people to natural elements in their surroundings with emphasis on trees and wooded areas or larger, open spaces such as parks (Smardon, 1988). Although this information is extremely valuable and serves to guide the research of horticulturists, it does not adequately address the types of plants, the uses of plants, or the active participation with plants that are important to members of the horticultural community or to the majority of the residents in urban and suburban communities.

In initiating the development of a baseline of information, we have identified disciplines from which to establish research partnerships and have acknowledged the lack of research focused on horticultural crops and activities.

The next step in implementation is to determine the need for this research and the interest in conducting it. There is some debate as to whether the information to be gained from conducting people-plant research is so self-evident as to be redundant and thus not worth gathering. Papers presented in this symposium refute this line of reasoning, however, and establish the value of such research. For example, Mike Evan's study shows that for an initial $1 million investment and annual maintenance costs of $1.2 million, Opryland has achieved a net gain of $7 million per year from the rooms surrounding its atrium (Evans, 1990). This is certainly a strong argument for developers to install and maintain atriums.

One of Roger Ulrich's classic studies documented that gall bladder patients with a view of vegetation went home earlier and required less medication than similar patients with a view of only a wall (Ulrich, 1984b). This study and related studies (Moore, 1982; West, 1985; Jesse et al., 1986) have just begun to explore the impact of plants on human health and hospital recovery factors. Related research into the impact of the hospital landscape and floral gifts, possibly even the integration of atriums or patient "sun spaces" in the hospital, could provide information that would have an effect on hospital design, insurance payment, horticultural producers, and, most importantly, patients.

The changes in human behavior, physiological responses, economic impact, perceived value, and similar quantified data collected by researchers reporting at this symposium confirm the need for further research. In addition, the symposium documents the extent of interest in this area. With 160 participants representing the United States, Canada, Germany, New Zealand, and Japan, there can be no doubt that this is a viable field of research for members of the horticulture community and other disciplines. The diversity of participants is also significant: academians (that is, researchers and educators) constitute approximately 45% of the participants, commercial horticulturists represent 18%, arboreta and botanic garden personnel comprise 9%, horticultural therapists make up 12%, and approximately 16% of the participants are from related fields (the psychologists, landscape architects, physicians, and artists who are our counterparts in interdisciplinary research).

The importance of this type of research to the horticulture community is indicated by the support both in endorsements and funds received for this symposium. A recent issue of the American Society for Horticulture Science (ASHS) newsletter (1989) cites "Enhancement of human living environment" as the number 7 research priority as ranked by the Science Priorities Committee. Fifty ASHS members indicated this as their primary area of activity and an additional 39 indicated it as a secondary activity. These researchers are not necessarily addressing the psycho-social impact of plants on people in their research, but they represent a cadre of researchers with related interests who may be able to expand their research to include people-plant issues.

As further substantiation of the validity of this area of research, we can examine the idea that the 1990s promise to be the decade of the environment. People are deeply concerned about the state of the environment on an international, national, and local level. They want to take action to secure it from degradation and enhance its impact on life quality (aesthetically, economically, physically, and psychologically).

On this Earth Day weekend, we are surrounded with numerous indications that the 1990s will be focused on the environment in activities ranging from legislative action to volunteer projects. Global Re-Leaf is an example of the many programs directed toward taking action to save our environment. On television commercials and in other advertising, there is increased visibility of plants as advertisers use horticulture to sell other products. McDonald's current giveaways are Peanut cartoon characters "gardening," Wall Street is now in the arena with environmental stocks, and there is even an MTV video on growing a sunflower (Tears for Fears, "Sowing the Seeds of Love").

Much of the action to improve the environment can and must be taken on a local level and has a horticultural element. Examples include composting to reduce the load on our landfills, planting to increase the biomass with its impact on air quality and energy consumption, and gardening to reduce mental fatigue. This revived interest in the environment combined with health and nutrition consciousness has helped to make horticulture one of the fastest growing segments of agriculture (U.S. Dept. Ag., 1988). At the farm or producer level, agricultural crop receipts total 72.6 billion; 35% (25.6 billion) are from horticultural crops, including vegetable, fruit, and greenhouse/nursery crops (Johnson, 1990). This does not take into consideration the horticultural service industry or the impact on property value, tourism, etc. Garden center sales grew at twice the rate of all other retail trades between 1982 and 1987 (U.S. Dept. Commerce, 1990).

It might appear that there would be no benefit for members of the horticulture community to provide quantified research regarding the people-plant interaction, since there is a certain amount of research under way. That is hardly the case, however. A drive through most of America's commercial, public, or residential districts will quickly allow you to identify major areas where life quality would be greatly enhanced by an investment in better landscape or the inclusion of community, school, or corporate gardens. The current projects and research substantiate the need for more information with increased emphasis on the enhancement of the man-made environment through horticulture. (Eberbach, 1990; Sommers, 1984; Edwards and Dammann, 1985).

The market for flowers exceeds the domestic supply and imports have become an increasingly large part of the floriculture economy. K-Mart is America's number 1 nursery retailer (Brantwood, 1989) because it takes advantage of people's demand for convenience, desire for plants, and practice of impulse purchases. Despite these positive signs of increased market, studies indicate that the United States' per capita horticultural/floral purchases are less than half of Europe's (U.S. Dept. Ag., 1988). A better understanding of people's physiological, psychological, and social response to potted plants and cut flowers would serve as an excellent marketing tool while also providing information to help enhance life quality.

The Opryland Hotel reported tremendous return on the investment in its atrium; however, interior designers are shifting the look of restaurants, hotel lobbies, and even the layouts in home and lifestyle magazines away from interior plants. At the same time, they are increasingly featuring exotic cut flowers. Is this shift a whim of the designers? How does it influence the guests in hotels? What impact will it have on the horticulture industry?

The Foliage for Clean Air Council is poised to conduct a massive marketing promotion on the value of interior plants for improved air quality based on controlled atmospheric studies funded by NASA. Building consultants are concurrently posing questions as to the negative impact of plants (i.e., high humidity causing damage to buildings, especially in the southeastern part of the United States; off-gassing of containers adding more pollutants to the air than the plants remove; foliar fertilizers introducing nitrogen into the air; and organic mulches increasing the presence of bacteria). These factors, both "pro" and "con," do impact human comfort and, ultimately, job satisfaction and productivity; however, they do not take into account the psychological influence of plants. For example, are people more relaxed and comfortable in a room with plants? Initial studies indicate that they are (Askawa et al., 1984; Laviana et al., 1983). Do employee morale and worker productivity actually increase due to the proximity of plants, as some claim? If so, does the increase offset the cost of the plants and maintenance? Cost/benefit analysis is usually the bottom line with corporate America, and additional data are essential for accurate analysis.

New arboreta and botanic gardens are being developed across the country. In Virginia alone, five have been established in the last ten years. Nonetheless, budgets are tight and these facilities often experience great difficulty in obtaining funds. Private gardens such as Longwood can not survive on admission tickets, but instead rely on generous endowments. This new expansion in arboreta and botanic gardens indicates the public interest in plants, but the low level of funding underlines the lack of understanding regarding the role of plants and public gardens, which might be considered as "plant cultural centers" or "plant museums" in human culture, creativity, and well-being.

Many people are extremely interested and willing to take action to protect the environment on both a global and a community or backyard scale. Observation shows that the presence of plants changes people's actions and feelings about themselves and their neighbors (Lewis et al., 1972). Research has shown that the effect of plants even extends to fatigue levels (Kaplan, 1987), stress (Honeyman, 1987), physiological processes (Ulrich and Simons, 1986), and other attributes of life quality (Wise et al., 1990).

The stage is set for implementing a research initiative that can have long-lasting impact. The questions are no longer: Is anyone involved in legitimate research? Is there a need for more research? The questions are now: Will the members of the horticulture community be involved in this research initiative? And, if so, how can this come about?

For the purpose of this paper, we identify the members of the horticulture community as those individuals who have a committed interest in horticultural crops and services, usually as their primary means of financial support. We also include, however, those individuals who participate in horticulture as a strong avocation and contribute to society through volunteer activities. Members of the horticulture community include, but are not limited to to following groups:

- academicians (universities, colleges, community colleges, American Society for Horticultural Science)

- arboreta and botanic garden personnel (American Association of Botanical Gardens and Arboreta)

- therapists (American Horticultural Therapy Association)

- industry (American Association of Nurserymen, Society of American Florists, Associated Landscape Contractors of America, other commercial producers, support industries—greenhouses, pots, equipment, etc.)

- amateurs and their association staff (Master Gardeners, National Gardening Association, American Horticultural Society)

- urban horticulturists (American Community Gardening Association, community gardening staff, city horticulturists, garden clubs, and others)

- communicators (Garden Writers Association of America, Extension service)

All of these individuals would benefit directly and indirectly by the application of research that develops a better understanding the people-plant interaction.

Based on representation at this meeting, I believe that members of the horticultural community are ready to become involved. On Sunday, 22 April 1990, a group of 30 leaders representing diverse areas in the horticulture community will have an agenda-setting session to consider our commitment and recommend how we should proceed. Reports of this session will be in the proceedings (see section entitled, *A Look to the Future*). The support from members of the horticulture community is important for researchers from related disciplines who wish to expand their research to include horticultural crops and activities. It opens the door to the development of research partnerships within horticulture departments, arboreta and botanic gardens, and other education/research-based facilities. It increases the options for sources of research money, supplies, and sites. This involvement of the horticultural community expands the number of receptive journals, trade, and amateur garden publications as places to report results.

I would like to take this opportunity to share with you some of my thoughts on implementing a research initiative. For this presentation I have identified three elements for establishing a comprehensive and effective program of research for what I have begun to refer to as Human Issues in Horticulture [2]:

- defining and conducting the research (this includes funding the research)
- communicating the findings
- applying the results

As in any research field, part of the research will be directed toward developing a basic level of knowledge and understanding, which may not have immediate or obvious application to improving life quality. This basic research will serve as the framework around which applied research can develop. Much of the research, however, must address issues of immediate concern. Researchers must also collect data for decision-makers, whose actions play a formidable role in life quality for people of all socioeconomic groups.

## IDENTIFYING AND CONDUCTING RESEARCH

For the greatest immediate and long-range impact, broad research agendas need to be established. This could be accomplished by special meetings or "think tanks" which include researchers, providers of funds, producers, and users. All of these groups need to be involved to achieve a balanced research agenda.

At the local research level, where specific projects are defined, a preliminary step must be the development of interdisciplinary partnerships. Researchers in horticulture departments, arboreta, and botanic gardens network extensively with other plant and soil scientists. Now we need to include social scientists and researchers from the humanities. Human Issues in Horticulture (HIH) crosses over many disciplines, thus lending itself to valuable interdisciplinary work. At Virginia Tech we have established the Interdisciplinary Research Team in Consumer Horticulture (IRTCH), a group of faculty, staff, and graduate students interested in horticulture's role in society and the roles of plants in history, culture, wellness, and behavior. With 19 members representing 12 departments, it has been possible to combine unique skills and to integrate HIH research into existing research programs. Currently, six cooperative interdisciplinary projects are underway.

The interdisciplinary research team approach has many benefits:

- It offers opportunities for cross-fertilization of ideas as individuals in seemingly unrelated areas recognize shared interests.

- It avoids the negative perception of work being *cross*-disciplinary, wherein individuals in one discipline expand their research into another discipline without benefit of interaction with colleagues.

- It serves as a focus for other activities. For example, the IRTCH held a campus seminar and workshop on people-plant issues on 31 October 1989 that featured Rachel and Stephen Kaplan, Charles Lewis, Chuck Richman (Executive Director of AHTA), Charles Dunn (president of a prominent horticultural marketing firm), and other industry representatives. The University Provost opened the event by emphasizing the importance of interdisciplinary cooperation, especially in the study of people-plant interaction. The seminar was well-attended by faculty, staff, and students from many departments, and the afternoon discussion session was dynamic and productive. The campus seminar served as a preliminary experience in preparation for this symposium, and many members of the IRTCH have contributed to this event.

As we have learned by initiating research projects at Virginia Tech, it is possible to address Human Issues in Horticulture within the context of on-going research agendas. Table 1 shows the traditional horticulture research areas that could easily be expanded and explored with researchers from other disciplines. Table 2 shows additional expansions of existing research from the perspectives of other disciplines.

**Table 1.** Traditional areas of horticulture research that could be pursued cooperatively with researchers in other disciplines.

| Area | HIH Issue | Related Discipline |
|---|---|---|
| Quality nursery stock | Consumer perceptions and preferences | Psychology |
| Cut flower market | Influence of table flowers on customer satisfaction | Hotel/restaurant management |
|  | Consumer perception of advertisements | Communication and marketing |
| Wildflowers along highway and corporate plantings | Tourist/employee/neighbor perception and acceptance | Behavioral psychology |
| Effectiveness of different pot mulches | Influence on air quality and bacterial growth in interiorscapes | Building construction |
| Landscaping with bedding plants | Influence on litter control | Behavioral psychology |
|  | Impact on community image by company considering a move | Business/Psychology |

**Table 2.** Areas of research in other disciplines that could be pursued cooperatively with horticulture researchers.

| Area | HIH Issue | Related Discipline |
|---|---|---|
| Social support and risk behavior of university freshmen | Influence of plants on lifestyle | Behavioral psychology |
| Optimum amenities in hotel guest rooms | Role of cut flowers, foliage plants | Hotel/restaurant management |
| Regional growth management and regional connectedness | People-plant viewpoint | Urban affairs and planning |
| Computer-based interactive video systems | User preferences for horticultural crops and designs | Computer science/extension |

A lack of understanding of the research methodology used by the social scientist can severely limit the researcher trained in the plant sciences, a mode that lends itself to much greater control and manipulation. Schroeder (1986) clearly defines two ways of understanding the psychological value of urban trees. These two methods apply to much of the research conducted in Human Issues in Horticulture. Schroeder states:

> The first way, which I call the measurement approach, is based on a rational, scientific world view. It records subjective preferences and values on numerical scales in a manner comparable to the measurement of physical benefits such as air and water quality. Examples of this approach include rating scales of public perceptions, measurement of physiological responses such as pulse and brain waves, and economic estimates of willingness to pay.
>
> The second way of understanding psychological value asks about the meaning or significance of trees for people. This approach recognizes the importance of emotion, imagination, and intuition in people's experience of the natural world. It does not seek to quantify value, but to describe how people interpret their surroundings relative to their own experience.

Examples of the second approach would be focus groups in which individuals discuss specific issues relevant to the research question. The transcripts of these discussions are analyzed for ideas and meanings that are repeatedly brought up by various members of the group, thus indicating a significant agreement among group members as to their value. Schroeder goes on to explain:

> To scientists and managers, trained in the school of rational decision making, the measurement approach may appear to be more practical. If, however, we do not understand people's deeply held emotional and intuitive values, we may fail to realize the most significant benefits of trees in the urban environment. A combination of both approaches will provide the most useful and complete understanding of the value of urban trees.

Likewise, a combination of approaches is important in understanding the value of any horticultural crop or activity.

Tools for collecting research data range from surveys to interactive computers that monitor physiological changes. Accessing these tools effectively means that researchers from the horticultural community must learn new skills and work closely with individuals from other research areas.

The ease with which we were able to get members for the IRTCH may indicate the degree of interest in horticulture held by the public at large. Many other professionals recognize the relevance of this research to their work and the value of becoming involved in this unique and emerging area.

## COMMUNICATING THE FINDINGS

One of the complaints heard about research is that it lies dormant in scientific journals or sometimes as an unpublished Master's thesis or PhD dissertation awaiting discovery by another PhD or Master's candidate. Communicating the results of research must be planned as an integral part of a comprehensive research initiative. Provision should even be made for communicating negative or incomplete results, as they can serve as guides to other researchers. Particularly important is establishing a communication network to ensure prompt reporting to the users of the information.

There are several levels of communication of research information. Each has a valid place in the communication scheme. These levels of communication do not have to be sequential or mutually exclusive.

### Peers in Academia

Researchers are expected to publish in refereed journals with an established peer review system. Certainly at universities there is a legitimacy to this system because tenure and promotions are determined by a committee of colleagues from diverse disciplines who share no subject matter expertise and are often not actually qualified to comment on the researcher's expertise. Instead they rely heavily on the productivity of peer review research articles. There are innumerable drawbacks to this system. The three most significant are that it discourages applied research, it provides little incentive to communicate beyond peers, and it tends to delay the distribution and use of the research data.

Applied research is considered less prestigious than basic or theoretical research. As the number of researchers increase, the scientific journals increase their own prestige by publishing less applied research, thus restricting the outlets for this information to non-refereed publications that contribute little to the prestige or paycheck of the researcher. We in the horticultural community are fortunate to have several refereed journals that are responsive to publishing quality applied research and are interested in articles relevant to people-plant interaction. These include *HortScience* and *HortTechnology* from the American Society for Horticultural Science; *Journal of Environmental Horticulture* from the Horticultural Research Institute; and the *Journal of Horticultural Therapy* from the American Horticultural Therapy Association. In addition, there are journals from allied fields and other professional disciplines that are potential outlets for research articles. These include *Landscape Planning, Urban Ecology, Landscape, Environment and Behavior, Journal of Environmental Psychology, Landscape and Urban Planning, Building Science, Journal of Environmental Systems, Journal of Leisure Research, Journal of Arboriculture, Journal of Architectural and Planning Research, Landscape Research, Journal of Social Issues, Population and Behavior.*

The discrepancy in reward for publication of refereed articles versus nonrefereed is a strong disincentive to spend the time necessary to publish on other levels. As Paul Larsen (1972) warned, however, when we only publish for our colleagues, our work does little more than further the disciplines. One might say that we are stockpiling knowledge when we need to be investing it in the public for growth.

The time from completion of a research project to submission of an article is not infrequently one to two years. From the time a journal receives the article until it is printed, another year or more may elapse as the article goes through various editors, reviewers, and rewrites. An unknown number of years may elapse before the published research is connected with other similar research to observe patterns that have application.

A well-planned network of researchers and communicators can shorten this lengthy process and expedite the application of research findings. This, in turn, tends to increase research funding, increase the number of researchers, and increase the quantity and quality of knowledge gained.

## Professionals in the Business World

Business professionals who produce or use plants are easily targeted through their trade publications and their conferences and trade shows (examples include commercial horticulture, developers, hotel/restaurant management). Communicating with these professionals requires a different organization and presentation of information—one that is often alien to the science-based researcher. Because of the difference in audience and the nature of the information shared, it is appropriate to provide well-documented research to the industry before it has been published in a referreed journal. This can reduce implementation time by several years. Nonetheless, the peer review process is integral to quality research—analogous to a self policing process in some professional fields. Peer review should not be neglected after publishing for commercial application.

## Pupils at All Levels

There is a strong tendency for researchers to focus their educational efforts on fledgling researchers in graduate school. Yet what children learn at a very early age influences the rest of their life, including career choice and how they make decisions. For long-range sustainability of the impact of People-Plant Interaction research, the educational process must reach grades K–12, community colleges, universities, and continuing education settings. Understanding the values and benefits of plants and thereby learning greater respect and enjoyment from them cannot start too soon. Nor should it be limited to individuals who have expressed an interest in a career working with plants. Students with career aspirations in areas as diverse as special education and business management can be positively influenced by plants and can apply information about people-plant interaction to their chosen career (e.g., special education, by using a garden as a focus and motivator to developing math and language concepts; business management, by using a garden as part of a corporate incentive program). A course on people-plant interaction should be part of the university-wide core curriculum from the College of Agriculture because of its clear application to all students.

## Public

The public is interested in this subject. The problem is translating the research data from the researchers' perspective to something to which the public can relate. Several factors would have to be addressed to increase significantly the availability of information on people-plant interaction to the public. The factors are generally linked to constraints placed on the researcher and include the following:

- lack of understanding of the essential nature of this type of communication

- lack of time to communicate beyond peers and perhaps professionals

- lack of professional recognition or support for this type of communication

- lack of appropriate communication skills to be understood clearly

- reticence to share the information beyond immediate colleagues, particularly when the research is in its infancy as is most horticulturally oriented people-plant interaction research.

The education of the public regarding the value of people-plant interaction is intrinsic to justification of conducting the research. If the goal of this research is to provide understanding and the increased use/effectiveness of horticulture as a means of enhancing life quality, then certainly people must be made aware of documented effects in order to gain maximum benefits. Analogies might be drawn to research on the importance of good nutrition to health, or exercise to physical fitness, and the extensive educational and informational programs that have been used to encourage those behaviors. Public education is integral to incorporation of new understanding into our lifestyles.

Lack of time and communication skills on the part of researchers plays a role in slowing down this level of communication. There are other professionals, however, whose careers are based on public education. These include Extension specialists with the Cooperative Extension Service; mass media professionals with newspapers, radio, and television, freelance writers who prepare magazine articles and books; and information officers at most universities.

Accessing these professionals is frequently difficult because of reticence to share information prior to the final "conclusive study" and/or because it may be perceived as an unseemly act of self-aggrandizement. Most researchers tend to pursue their research and let public education evolve when and how it will.

A network focusing on people-plant interaction that includes communication professionals can significantly alter the factors that slow the process of public education by keeping communicators appraised of findings on a regular basis.

## Politicians and Planners

Roger Ulrich (1984a) explains the need for communication at the level of politicans and planners.

Most of the people responsible for urban development—the politicians and the planners—probably agree that plants contribute to environmental quality. Yet in the cost-benefit analyses that so often shape development decisions—where the benefits, as well as the costs, must be demonstrated and quantified—the intuitive arguments in favor of plants carry little weight. As a result the politicians often give only lip service to plants and dismiss planting programs as unwarranted luxuries.

The paucity of urban plantings attests to a wide gap between our needs, which cannot be expressed in numbers or dollars, and our planning processes. Recently, however, researchers from several disciplines have begun investigating the benefits of contact with plants. The research to date, still relatively meager but growing, has already deepened our understanding of the positive experiences plants make possible and the needs they satisfy. Such studies, giving plants a measurable value, may eventually enable urban planners to give vegetation the high priority it deserves.

## APPLYING THE RESEARCH

Conducting research and then communicating the findings is inadequate to bring about changes rapidly enough to meet today's needs. A plan needs to be established and implemented to ensure that the skills and resources exist to put the research into action immediately.

Developing and implementing an action plan to ensure application of the research is not necessarily the role of the researchers, but they should be involved to share their knowledge and ideas. Likewise communicators, educators, and users of the research data could effectively work together to develop a plan.

The goals of a plan of action could include

A   *Altering* individual behavior
C   *Committing* to community well-being
T   *Teaching* a new role for horticulture
I   *Improving* horticultural practices
O   *Organizing* human culture/horticulture events
N   *Networking* for effectiveness and impact

### Altering Individual Behavior

Horticulture has been shown to be an effective tool for altering individual behavior in a number of situations, including

- therapy and rehabilitation for disabled and elderly persons

- decreased mental fatigue in the workplace

- improved recovery rate in health care facilities

- selecting a room with view of an atrium to increase vacationing pleasure.

A comprehensive plan of action for applying research to this area could include training programs, consulting services, and written guidelines to assist professionals whose work focuses on individuals in using horticulture to meet their goals. Some associations such as AHTA are beginning to try to provide some of these resources.

### Committing to Community Well-being

Horticulture holds the potential for having positive impact on many aspects of community well-being. Partners for Liveable Places is most noted for their efforts in urban developments. According to a Partners representative who spoke at the 1989 American

Association of Botanical Gardens and Arboreta (AABGA) annual meeting, plants are part of the community infrastructure (i.e., roads, sewers, bridges) that is essential to the communities' health. Plants are also one of the least expensive and fastest ways of enhancing the image of an area, sending out the message that this is a good place to live. A third asset of plants in urban settings is equitable access to natural environments for the many urban residents who cannot go to the countryside or parks.

Plants play a role in the economic development of many areas through their impact on tourism. Theme parks, golf courses, historic sites, and recreational facilities rely on the ambience created by plants as a factor in maintaining the interest of tourists. It is not known how significant this is; however, a recent Virginia Tech/National Gardening Association survey reports that 50.1% of people feel that "the flowers and plants at theme parks, historic sites, golf courses, and restaurants are important to (their) enjoyment of visiting there."

Charles Lewis spoke at this symposium on the value of gardens in intergenerational bonding, as everyone, from children to grandparents, works together toward a harvest. Horticulture in classrooms and on the school grounds provides opportunities to experience and understand nature more intimately than reading or observing. There is an increasingly widely held concept that children need experience with nature such as a garden can give them as part of their developmental process. Shalala and Vitullo-Martin (1989) maintain that the federal Department of Housing and Urban Development builds instant slums because HUD housing does not include community centers, landscapes, trees, yards, and sidewalks

As with altering individual behavior, committing to community well-being through horticulture will require the support of professionals who are knowledgeable about people-plant interaction and understand the care and culture of plants in an urban environment to provide training, consulting, and written support material. A number of groups are working toward implementation of research and observations in this area to enhance the community, including Partners for Liveable Places, National Gardening Association, and Philadelphia Green.

## Teaching for a New Role for Horticulture

For horticulturists to play a role in understanding and enhancing life quality through horticulture, there is a need for changes in courses and curriculum at the university level. Horticulture students need better preparation to work with people, whether as managers or as part of planning teams. Horticulturists will be needed in the conceptualization and planning for appropriate use of plants, if plants are to provide the desired benefits. Horticulture students likewise need to increase their communication skills. In the Information Age, being able to grow a plant will not be sufficient to ensure career success. The ability to transfer information to the consumer will be the ultimate test of success. Courses for nonmajors need to better focus on how the student will use the information, skills, and knowledge of horticulture to enhance personal life quality and to support career goals.

## Improving Horticultural Practices

The stress of urban and interior environments on plants demands changes in cultural practices and cultivars to maintain vigorous healthy plants. In addition, the recognition of the impact of some current cultural practices on human health and long-term environmental quality requires a new evaluation of cultural practices. As the focus shifts from production and maintenance of horticultural crops exclusively for economic gain to an understanding of the benefits and value to people, new resources should become available to address horticultural issues such as stress management, cultural practices for a sustainable landscape, and cultivar development and/or selection for minimum maintenance.

Research for appropriate techniques and plants for use in enhancing people-plant interaction must go hand in hand with research to understand and utilize people-plant interaction.

## Organizing Human Culture/Horticulture Events

One of the areas of human culture most neglected by social science and the humanities is the garden. Yet the garden and the plants of the garden play an essential role in our great religions, our art, our language, and our life rituals from birth, to marriage, to death. Plants provide inspiration for folklore, mythology, and literature. To understand our heritage, museum exhibits looking at the daily life of our ancestors should include descriptions of the time spent in the garden, the crops grown, the tools used, the seed saved. Food and clothes have not always come from the store, and too few children today realize that the source of these products is not a factory.

Arboreta and botanical gardens serve as an essential plant-based part of the culture of a community. A focus on the interactions between human culture (fine arts, music, and theater) and horticulture could open doors of greater understanding of the importance of plants in the diversity of human culture.

## Networking for Effectiveness and Impact

Networking has fallen into some disfavor as a buzzword of the 1980s. But the concept of a network as an organized way of linking with others to exchange information, advice, contacts, and support is critical to the success of any multidisciplinary, multifaceted project. Networking is an informal, though organized, way to work toward implementation of research findings. By linking the many existing efforts in research and application of people-plant interaction to integrate research findings, a support system is developed that has greater impact than the individual elements or projects that are part of the network. Although researchers often are not involved in the application of their findings to society, they can be an essential element of the network, providing clarification, consulting, and increased visibility for the information.

## CONCLUSION

Initiating a research program appears to be as simple as conducting a specific research project. But sustaining the research and achieving an impact is considerably more complex and involves the cooperation of individuals from many disciplines. Professionals in environmental psychology, urban forestry, and landscape architecture have been exploring the people-plant interaction for almost 20 years. They have built an important body of knowledge regarding this relationship. It is time that the members of the horticulture community work with them in this search for understanding.

Traditionally, the diverse associations in horticulture—American Association of Nurserymen, Society of American Florists, Associated Landscape Contractors of America, American Society for Horticultural Science, etc.—have not worked closely together to focus their energies on shared concerns. Instead, their actions have been based on the demands of the specific crop or service that they represent. I am reminded of the neighborhoods that Charles Lewis described that deteriorated because the neighbors failed to work together and did not realize what a tremendous impact directed, cooperative action could have on their environment and quality of life.

The horticultural community seems to be similar in its failure to work together, and some of the signs of deterioration are evident: difficulty in finding qualified workers, low student enrollment, competition from other leisure-time pursuits (most of which are much more expensive), failure of small businesses, large corporations selling out of horticulture, failure to address environmental quality issues adequately. By focusing on gardens, Charles Lewis reports, neighbors became empowered to change their community interaction and improve their environment. Perhaps this area of people-plant interaction can function in the same way as the garden to help the different facets of the horticulture community work together to the common good.

The members of the horticulture community can make an important contribution to the environmental movement of the 1990s, to human life quality, and to their own economic well-being. The extent and sustainability of that contribution will be determined by how well the horticulture community cooperates in working toward the common goal of understanding the critical interaction between people and plants.

## NOTES

[1]People-Plant Interaction—the wide array of human responses (mental, physical, and social) that occur as a result of both active and passive participation with plants. Because of the human benefits plants engender, this interaction implies stewardship of plants that are susceptible to man's impact on the environment.

[2]Human Issues in Horticulture research—research to understand the influences that plants have on humans beyond the traditional concepts of food, fiber, and shelter. This includes, but is not limited to environmental influences (e.g., removing pollutants from office air or increasing pollutants due to poor management practices), economic influences (e.g., increase value of property), physiological influences (e.g., restoration from mental fatigue through contact with nearby plants), psychological influences (e.g., enhanced self-esteem), sociological influences (e.g., intergenerational bonding), or therapeutic influences (e.g., prescriptive treatment for disabled).

## LITERATURE CITED

American Society for Horticultural Science. 1989. Research needs of horticultural science. ASHS Newsletter 5(6):9.

Asakawa, S. 1984. The effects of greenery on the feelings of residents towards neighborhoods. Journal of the Faculty of Agriculture, Hokkaido University 62:83–97.

Brantwood Horticultural Research. 1989. Retailer 100: Annual ranking of America's largest nursery retailers. Nursery Business Retailer, March.

Eberbach, C. 1990. School children: From recipients to participants. The Public Garden: Journal of the American Association of Botanical Gardens and Arboreta 5(2):14.

Edwards, S. and C. Dammann. 1985. Uncovering homeowners' hidden landscaping desires. American Nuseryman, Aug 15:117–123.

Evans, M. 1990. People and plants: A case study in the hotel industry. Paper presented at the symposium The Role of Horticulture in Human Well-Being and Social Development, Arlington, VA.

Honeyman, M. 1987. Vegetation and stress: A comparison study of varying amounts of vegetation in countryside and urban scenes. Unpublished Master's Thesis. Department of Landscape Architecture, Kansas State University, Manhattan.

Jesse, P., M. P. Strickland, J. D. Leeper, and C. J. Hudson. 1986. Nature experiences for hospitalized children. Children's Health Care 15(1):55–57.

Johnson, D. 1990. Personal interview with Doyle Johnson of the USDA Economic Research Service. 24 May.

Kaplan, S. 1987. Mental fatigue and the designed environment. Proceedings of the 18th Conference of the Environmental Design Research Association, Ottawa, Ontario.

Larsen, R. P. 1972. The challenge of communications. HortScience 7(5):436–437

Laviana, J. E. 1983. Plants as enhancers of the indoor environment. 27th Annual Meeting of the Factor Society, Norfolk, VA.

Lewis, C. A. 1972. Public housing gardens—landscapes for the soul, pp. 277–282. In: Landscapes for living, USDA Yearbook of Agriculture.

Moore, E. O. 1982. A prison environment's effect on health care service demands. Journal of Environmental Systems. 11(1):17–34.

Schroeder, H. W. 1986. Psychological value of urban trees: Measurement, meaning and imagination. Proceedings of the 3rd National Urban Forestry Conference. American Forestry Association, Washington, D.C.A

Shalala, D. E. and J. Vitullo-Martin. 1989. Rethinking the urban crisis: Proposals for a national urban agenda. Journal of the American Planning Association 55(1):3–13.

Smardon, R. C. 1988. Perception and aesthetics of the urban environment: Review og the role of vegetation. Landscape and Urban Planning 15:85–106

Sommers, L. 1984. Theory G: The employee gardening book. National Association for Gardening, Burlington, Vermont.

Ulrich, R. S. 1984a. The pychological benefits of plants. Garden, Nov./Dec.

Ulrich, R. S. 1984b. View through a window may influence recovery from surgery. Science 244:420–421.

Ulrich, R. S. and R. F. Simons. 1986. Recovery from stress during exposure to everyday outdoor environments. In: J. Wineman, R. Barnes, and C. Zimring (eds.). The costs of not knowing: Proceedings of the Seventeenth Annual Conference of the Environmental Design Research Association. Environmental Design Research Association, Washington, D.C.

U.S. Department of Agriculture. 1988. Commodity spotlight: Ornamental horticulture is booming. Agricultural Outlook AO143:12–14.

U.S. Department of Commerce. 1990. Census of Retail Trade 1982–1987. U.S. G.P.O., Washington, D.C.

West, M. J. 1985. Landscape views and stress response in the prison environment. Unpublished Master's Thesis. Department of Landscape Architecture, University of Washington, Seattle.

Wise, J. A., K. McConville, and N. Al-Sahhaf. 1990. Managing cultural diversity in orbiting habitats. In: Proceedings of Space '90, Albuquerque, New Mexico.

CHAPTER 37

# Chicago Botanic Garden Examines Its Social and Economic Role in the City of Chicago and Cook County

Sue Burd Brogdon

Director of Education, Chicago Botanic Garden

Historically, the social and economic value of cultural institutions to a community have been overlooked and undervalued. When budgets are tight and social problems press, cultural and educational institutions are often seen as being peripheral. In the last decade, however, policy makers and administrators have begun to reexamine the value that museums, zoos, and botanical gardens can provide to society.

## HISTORICAL OVERVIEW

The Chicago Botanic Garden is a contemporary garden owned by the Forest Preserve District of Cook County and managed by the Chicago Horticultural Society. In 1990, the Garden will be celebrating its 25th ground-breaking anniversary. During the past 25 years, the primary focus of the institution has been building gardens. Today there are 14 major display gardens, educational programs for a variety of audiences, and a fledging research program. Support comes not only from the county but also from corporations, foundations, and governmental agencies. The membership totals 14,000, and one half million visitors pass through the garden gate each year. Outreach programming targets Chicago Public School children, community gardening groups, and horticultural therapy programs for disabled audiences.

## THE CONTEXT

Cook County's population of 5.8 million people comprises a majority of the 8.5 million citizens in the greater metropolitan area. The county covers over 958 square miles; and the Garden is located at the northernmost boundary, 30 miles north of downtown Chicago, in a very affluent area.

The Garden receives almost 75% ($6M) of its annual operating budget from the county, but serves a small percentage of the citizens under its charge. Cook County's budget allocation supports three primary functions: the county hospital, sheriff's office and county jail, and 66,915 acres of forest preserve land managed by the Forest Preserve District of Cook County. Both the Chicago Botanic Garden and the Brookfield Zoo receive funding from the county under the auspices of the Forest Preserve District.

## THE ISSUES

As the Garden moves into its second quarter of a century as a major cultural institution in the county, it must position itself to meet the growing social, educational, and economic needs of all segments of the community. The Garden cannot sit back and be a spectator to the activities and issues that influence our lives. We must participate in developing programs that are integral to the needs of our constituencies. Before we can begin to formulate programs and build partnerships, we need to wrestle with the issues impacting our society. We live in a multifaceted and culturally diverse society. Are our programs and services representing the needs of these people? Chicago is reputed to have one of the worst school systems in America. What is our responsibility as an educational institution? Can we develop programs to help combat this problem? Environmental issues such as pollution, loss of open land, and urban blight cover the front of our daily newspapers. Should the Botanic Garden take a leadership role in adressing these problems? A myriad of questions come to mind; the administrator must determine which issues the institution should embrace to further its mission.

## THE PARTNERSHIP

Since these issues are complex and frequently interrelated, the Chicago Botanic Garden sought the counsel of Partners for Livable Places (Partners), a not-for-profit organization, to help formulate a plan of action to assess and analyze the role of the Garden in relation to the economic and social needs of the City of Chicago and Cook County.

Partners promotes the economic health of communities through better planning, management, and design. Their mission is to "encourage a greater public consciousness of our physical surroundings and their economic and social consequences." Partners emphasizes the "process of partnership, the importance of local initiatives and the value of cooperative learning between public and private sectors."

In 1987, Partners initiated a national demonstration project, Shaping Growth in American Communities, which was developed to meet the challenges of unprecedented change in communities throughout North America. Thirty-six organizations will be participating in this project, designed to develop public policy statements to address issues and to construct regional models to guide community and institutional growth in the 1990s.

In 1988, Partners approached the American Association of Museums (AAM), a Washington, D.C.-based, not-for-profit association representing the needs of all of America's museums, to determine whether AAM would like to be a collaborator and a cosponsor of this national demonstration project. Partners and AAM identified three central issues that relate to the changing role of museums in their communities: (1) the need to increase institutional accessibility to all people and to develop programs responsive to these audiences; (2) the need to focus programming endeavors on lesser served members of the community such as minority and lower income audiences; and (3) the need to establish collaborative ventures with other civic and social organizations.

The Garden approached Partners at an opportune time; Partners and AAM were just beginning to select museums around the United States. Both were eager to utilize a "living" or nontraditional museum as a test site for the Chicago cultural community.

## FUNDING

The Chicago Botanic Garden in collaboration with Partners and the AAM jointly authored a grant proposal. The Chicago Community Trust and The Robert R. McCormick Charitable Trust each contributed $40,000 to support the study. Notification of funding was received in the fall of 1989.

## THE METHODOLOGY

Last November, senior Garden staff members and representatives from Partners met with AAM to review the project proposal, develop a research plan including the formation of an Advisory Committee and focus groups, and outline a timetable. From the onset of the project, staff determined that it was essential to involve Board members in the process from its inception, if the Botanic Garden were indeed going to embrace this initiative. The Chairman of the Garden's Board of Directors appointed an Urban Task Force Committee, cochaired by two Board members who have strong political and business ties in the city.

During the winter months, Partners began researching the issues of importance to Chicago and Cook County in the future. Partners began an exhaustive search for reports dealing with educational, political, social, and business issues and spoke to many influential people in the city. Concurrent with this, Partners began surveying public gardens for information on innovative and exemplary community outreach programs.

In January, a meeting was convened at the Garden to review the status of the findings to date, establish an Advisory Committee, and determine which people who would be interviewed as part of the focus groups. The role, purpose, and composition of the Advisory Committee was formulated. It was determined that in addition to having two Board Members from the Garden, a cross-section of city leaders drawn from a broad community base including both the private and public sectors would be beneficial to the project. The Advisory Committee would serve two primary purposes: (1) to assist Partners, AAM, and the Garden by guiding research and program development and providing feedback on program development, and (2) to help solve problems and establish city, community, and organizational networks.

It was also decided that the most realistic way of gathering data about Chicago and involving Chicago's leaders in the process would be through the use of focus groups. Four focus groups are being planned, representing the voices of Chicago's business, educational and cultural, political and social communities. Each focus group session, planned for mid-April, will be led by a professional facilitator. The facilitator will engage groups of 12 to 15 people in a dialogue about the issues, opportunities, and needs of the greater Chicago area. These conversations will be recorded and transcribed. Staff from Partners will then prepare a written summary report. Partners will also present an interim report to the entire Board of Directors at an April meeting.

In May, Garden staff, Partners, and AAM will meet to review the results of the project research. Partners will begin analysis and assessment of the materials gathered from the preliminary research, focus group meetings, and previous meetings with Garden staff.

Following this, a position paper will be prepared for review by Garden staff and the Advisory Committee. Changes will be made based on recommendations from these two groups. In October, a project report will be presented to the Garden's entire Board of Directors. It is the intent of the Garden's administration to use this data as a benchmark to evaluate existing outreach programming endeavors and as a catalyst for launching new program initiatives.

## SUMMARY

All not-for-profit institutions have finite resources and, in many cases, are under public scrutiny to ensure that they are serving all people. If cultural institutions are to remain strong and viable in the 1990s, they must take risks, ask difficult questions, and be responsive to the needs of those audiences they purport to serve.

CHAPTER 38

# National Survey of Attitudes Toward Plants and Gardening

Bruce Butterfield

Research Director, National Gardening Association, Burlington, Vermont

Diane Relf

Associate Professor of Horticulture, Virginia Polytechnic Institute and State University

Each year, the National Gardening Association contracts the Gallup Organization, Inc. to conduct a consumer market research study called the National Gardening Survey. The purpose of this survey is to measure the types of lawn and garden activities that U.S. households participate in, how much money consumers spend annually on their lawns and gardens, the types of gardening products that were purchased, the demographics of U.S. households involved in lawn and garden activities, and a wide range of other lawn- and garden-related topics. The results of the National Gardening Survey are based on face-to-face, in-home, personal interviews with a representative national probability sample of more than 2000 households in all 50 states. The results of this survey have a sampling error of plus or minus three percentage points of the results that would have been obtained had the entire U.S. household population been interviewed.

For the first time, in 1989, a question was included in the National Gardening Survey that asked people if they agreed with a number of statements about the value and importance they place on plants and gardening. Survey respondents were handed an exhibit card with seven statements to read and were asked: "Which, if any, of the statements listed on this card about the value and importance of plants do you agree with? (Please call off the number of each of these statements that you agree with, you can choose more than one.)" The statements represented a range of attitudes towards plants and gardening from marginally positive to extremely positive.

The statements were as follows:

- Trees and flowers in a city are not important beyond their beauty or pleasing appearance.
- One of the most satisfying aspects of gardening is the peace and tranquility it brings.
- The flowers and plants at theme parks, historic sites, golf courses, and restaurants are important to my enjoyment of visiting there.
- Well-maintained landscapes and street plantings offset the loss of nearby natural areas to development.
- Gardening gives me a sense of control over my environment.
- Being around plants makes me feel calmer and more relaxed.
- The natural world is essential to my well being.

A total of 96% of all those interviewed agreed with one of more of the above statements. Only 1% of respondents said "none" and 3% of respondents indicated "don't know." Multiple responses totaled 242%, which indicates that people agreed with an average of just over two statements each from the list supplied. Table 1 shows the results for each individual statement in terms of the percentage and number of all U.S. households that agreed with that statement.

In general, higher-than-average ratings than the overall U.S. total for the above statements were seen in households of those aged 35 to 54, those with college educations, professional or business occupations, annual incomes of $40,000 and over, households in the West, and married households. U.S. households that participated in one or more types of gardening in 1989 also rated the importance of plants and gardening significantly higher than the national average. In 1989, 75% of the 92.8 million households in America, or an estimated 70 million households, participated in one or more types of indoor and outdoor lawn and garden activity and spent an estimated total of $16.3 billion.

**Table 1.** National Gardening Survey, 1989, selected statements and responses.

| Statement about the importance of plants and gardening to you | All U.S. households | |
|---|---|---|
| | % | Million |
| The flowers and plants at theme parks historic sites, golf courses, and restaurants are important to my enjoyment of visiting there. | 50.1 | 46.5 |
| The natural world is essential to my well being. | 46.0 | 42.7 |
| Being around plants makes me feel calmer and more relaxed. | 40.0 | 37.1 |
| One of the most satisfying aspects of gardening is the peace and tranquility it brings. | 37.1 | 34.4 |
| Well-maintained landscapes and street plantings offset the loss of nearby natural areas to development. | 33.1 | 30.7 |
| Gardening gives me a sense of control over my environment. | 23.5 | 21.8 |
| Trees and flowers in a city are not important beyond their beauty or pleasing appearance. | 12.0 | 11.1 |

CHAPTER 39

# An Extension Approach
# to Implement Research Results
# in the Flowering Plant Industry

Kevin L. Grueber

Assistant Professor of Horticulture, Virginia Polytechnic Institute and State University

## INTRODUCTION

Research results and concepts that relate human behavior and plants are significant within the scientific community because they substantiate what was previously mere dogma. However important the scientific facts may be, they are fruitless in real-world application unless they are translated and disseminated to the audiences that might consequently "profit"; specifically, the plant production and distribution industries and the consumer. An approach similar to that used in the Extension system is one possible mechanism to ensure that such information is communicated to these audiences.

In the Extension education and communication process, research results and information are collected, interpreted, and then disseminated to the appropriate audience. Because Extension personnel repeatedly employ this process of information transfer, they may be used to the advantage of the scientist researching the roles that plants play in human behavior and well-being. Since many scientists may not be familiar with the Extension Service, an understanding of both the Extension system and the flowering plant industry may help to ensure that applicable and appropriate research results are properly distributed.

## THE FLOWERING PLANT INDUSTRY

The floriculture industry is composed of producers (growers), wholesalers, and retailers. The crops involved may be flowering potted plants, bedding plants, cut flowers, or foliage. The grower typically provides a finished plant product, which is sold to a wholesaler who, in turn, sells the product to a retailer. The exact sequence may vary with the plant product and existing marketing channels. Most products follow this marketing path even though all the links in the chain may be the same person or business. Each one of these links in the marketing chain can potentially benefit by research on people-plant interactions and can be effectively reached with proper communication methods.

Providing assistance to the grower usually takes one of two tacks: technical assistance pertinent to production (e.g., cultural recommendations, problem solving, etc.); or business-related expertise. Most well-established greenhouse operations tend to be rather self-reliant regarding production, and Extension specialists tend to be more familiar with production aspects. The more difficult aspect of meeting grower needs tends to be in business-related matters, particularly in regards to market development and expansion. Extension specialists frequently hear statements such as, "My business is doing fine but could be better if only I had help with marketing."

Marketing, to the grower, may take several forms but could be basically defined as identifying and expanding both existing and new sales markets. For example, the average producer grows poinsettias for the period from Thanksgiving to Christmas. Even though national sales continue to increase yearly, many growers find that they are unable to sell their entire output. If, for example, basic scientific research implicated poinsettias as reducing the effects of holiday depression, the existing market potential could be expanded through consumer education and consumer-directed advertisement.

Such an expanded market would, presumably, benefit all participants; the producer because a greater volume of product would be produced and sold, the wholesaler and retailer because of the greater volume, and, ultimately, the "new" consumers who buy the product for its mental health value. Many of the research projects that investigate the roles of plants in human well-being could also result in the development of new markets for both existing and new plant products.

## CHANGES IN MARKETING AND MARKET CHANNELS

Traditionally, the availability of plant products in the United States has have been determined by the retailer, who predicts what consumer demand is or will be and then orders plant material to be produced. This process is, in many cases, being modified as a result of increased consumer awareness and a greater understanding and willingness of retailers to offer products that meet customer preferences. Research on people-plant interactions can be communicated to both the retailer and the consumer to increase sales and customer satisfaction and, ultimately, to benefit the plant production industries while increasing quality of life for the consumer.

## DESCRIPTION OF THE EXTENSION SYSTEM

The Extension system is typically composed of units (generally by county or city) that house agents who work directly with consumers and industry. The agents often rely on specialists (located regionally or, more typically, at land-grant institutions) for technical information in an area of expertise such as vegetable production, turf management, or greenhouse production.

In a general sense, the specialist collects pertinent information, research results, and scientific information and then passes the information on to the agents. Both the specialist and agent interpret this information and translate it into a vernacular accessible to the audience. It is the responsibility of both specialist and agent to select an appropriate medium or vehicle to reach and communicate with the target audience effectively. The target audience may be industry (production or marketing) or consumers.

## INFORMATION TRANSFER BY THE EXTENSION SYSTEM

The agent and specialist are responsible for employing a medium and style of communication that are appropriate for the intended audience. Technical bulletins, newsletters, articles in trade publications, and presentations at meetings are traditionally used for reaching the production and retail industries. Although the agent may assist with this process, the specialist is the primary user of these communication vehicles. Therefore, the researcher would be more likely to enlist the assistance of the Extension specialist when attempting to communicate with the plant production industries.

For the type of research presented and discussed at this symposium, the most likely target audience for research results would be the retail industry, including garden centers, florists, and retail greenhouse operations. These businesses have their own organizations, trade magazines, and other existing vehicles for communication of research results, information, and ideas. The retailer and the trade associations are in an effective position to pass the pertinent information on to consumers through advertisement and other marketing tools.

Aside from the Extension service, most universities (and even some academic units) employ information officers who are responsible for communicating the "news" of the researcher to appropriate audiences through news releases and publications. Such services can be used to reach both industry and consumer audiences.

Communication with the general public tends to require greater creativity, technical skill, and use of more diverse media vehicles than communication with industry. Mass media, audio-visual, and popular press are often used as vehicles for reaching the largest audience and achieving the greatest impact. Although many Extension specialists and agents effectively use these methods, many, unfortunately, do not. To be truly effective, Extension personnel need to become communications experts.

The mass media, popular press, and consumer-oriented media are underutilized in academe. This underutilization could be due to the faculties' lack of recognition for such efforts on the part of academic administration, aversion to publish in consumer-oriented media, or familiarity and expertise with such information dissemination practices and procedures. The increasing use of sensationalism and exaggeration in mass media may make researchers hesitant to use such media for fear that their research results may be misinterpreted or exaggerated. Nonetheless, very effective communication tools would thus be overlooked.

## CONCLUSIONS

The results of research on the effects of plants on human behavior and well-being have significance beyond the scientific community. The general public and the industries involved with plant production and sales can tangibly benefit by such research. In order to accomplish increased sales and profits in the plant industries, however, research results must reach the appropriate audiences in an understandable, usable form. By using communication methods such as those used by the Extension Service, and by using appropriate communication media and vehicles, research results can be properly collected, translated, and disseminated to the target audiences that will most greatly benefit.

# CHAPTER 40

# Beyond Romanticism: The Significance of Plants as Form in the History of Art

---

Rhonda Roland Shearer

Sculptor, New York City

## INTRODUCTION

When we think of the role of plants in art we contemplate a variety of images: the Hudson River School of romantic landscapes, Georgia O'Keefe's close-up paintings of flowers, still lives of the impressionists and post-impressionists. I am going to ask you to clear these images from your mind. We are going to look at plants not as beautiful, sentimental, or decorative objects, but as universal forms whose very structure offers a window into the underlying vital principles of nature itself.

## LEONARDO DA VINCI

Leonardo was the first artist in the history of art to truly observe nature beyond the symbolic, as illustrated in his *Madonna of the Rocks*, Louvre version, done in 1482. Leonardo's rendered plants were more than accessories to painted scenes depicting human beings. His care and study of plants is clearly indicated by his artistry of interpretation and their surprising botanical correctness, as exemplified by *Lilies*, drawn in 1508, from the Windsor Collection.

While his contemporaries stayed closed up in their studios using propped-up broccoli stalks as models for trees, Leonardo went out into nature, making frequent trips to the Vatican Gardens and studying the botanical books available to him. He observed the most minute morphological details. Leonardo describes trees more than any other subject in his notebooks, with over 100 passages of detailed notations on leaf phyllotaxis and distribution, branching growth variations, and the effect of light and terrain. Although only a handful remain in existence today, Leonardo had made hundreds of sketches of plants and flowers from his observations of the universal structures around him.

Life to Leonardo was like the infinite variety of music; however plentiful, all could be reduced to seven notes or twelve tones. Leonardo discovered that nature could be reduced to several fundamental shapes; the spiral, the branching pattern, the meander.

These structures were seen by Leonardo as fundamental truths in nature. The spiral and branching patterns he found in analogy everywhere. He would see spirals in leaves winding around their stems, spirals in human hair tendrils and water vortices. He even made a drawing of a calf fetus in utero shown as a spiral pattern.

The branching germination of a peach seed he saw as analogous to the vascularization of the heart, as well as the branching of trees and river systems.

Although his renderings of plant forms evoke a sense of beauty, Leonardo was not seeking beauty, but truth, through these natural structures. Stated best in Leonardo's own words: "Come o man to see the miracles that such studies will disclose in nature."

## PIET MONDRIAN

Like Leonardo, Mondrian, the father of geometric abstraction, sought discovery of nature's most fundamental principles. To look at *Composition No. 1 with Red, Yellow and Blue*, a geometrically gridded painting by Mondrian (1921), one would hardly guess the importance of plants in Mondrian's development. Upon investigation, however, one finds that plants, and specifically trees, are the single most important subject matter for his evolvement into abstraction.

Unlike many artists, Mondrian was never sympathetic with the human figure. *Chrysanthemums*, painted in 1909, is only one of many flower pictures Mondrian created early in his career. Granted, many of these were painted for commercial reasons. Nevertheless, it is clear that Mondrian rejected the figure as his muse and sought out the hidden dynamic in plants. *Dying Chrysanthemums* (1908) shows an almost spiritual transformation—the breaking down of the flower's form to reveal its essence.

The transmutation of natural form from realism to geometric abstraction is the signature of Mondrian. Looking at Mondrian's work in series, one observes a step-by-step evolvement, a gradual dematerialization of matter, which can be seen specifically in his series of tree works: *The Red Tree* (1908), *The Grey Tree* (1912), *Flowering Apple Tree* (1912), *Composition No. 3* (1912–1913), and *Oval Composition (Trees)* (1913). These paintings show how dramatically Mondrian evolved from realistic trees to pure geometric abstraction, *Composition* (1922) serving as an example.

This clear evolvement of Mondrian's vision, in which appeared the "essential" geometry of trees, is in keeping with his intensive study of Madame Blavatsky's and P. D. Ouspensky's writings. Madame Blavatsky taught that, although veiled, the true essence of the world was geometric. To quote Madame Blavatsky: "God geometrizes: dots, lines, triangles, cubes and finally spheres, why or how? Because nature geometrizes universally in all of her manifestations."

Ouspensky suggested in his 1912 book, *A New Model of the Universe*, that a leafless tree was the closest experience we could have in three dimensions to a fourth dimensional reality. Illustrating this point was a drawing captioned *Tree Diagram of the Fourth Dimension*, which looks very similar to Mondrian's *Grey Tree*, also done in 1912. Through his work, Mondrian lifts our vision of trees above sentimentality and enables us to view their structures as vehicles for what Mondrian envisioned as a higher reality.

A creative dynamic was generated when Mondrian imposed horizontal and vertical lines on tree structures in his search for the "universal." Nature showed a striking dissimilarity—rendered tree branches resisted Mondrian's lines that were carved within its structure, and eventually became overpowered by this Euclidean system. This dynamic process in Mondrian's work was an important reflection of his time. The conflict between nature and technology was in central focus during the early 20th century.

Euclidean geometry's point-line-plane was a metaphor for what was thought to be the triumph of man over nature with technology. Nature, although universal in world presence, did not seem to express universality in its structures. Its forms were unmatched and indescribable by the perfection of Euclidean geometries. Nature's multifarious forms—clouds, mountains, trees—were too complex and irregular or, to quote Mondrian, "too random and capricious" to be described by point-line-plane. Mondrian decided that this lack of ability to describe nature by Euclidean geometry illustrated man's superiority over nature. "Nature" Mondrian stated, "must be subdued by man for greater technology."

## FRACTAL GEOMETRY

Contemporary views of nature have been radically altered from Mondrian's time. The utopian hope of technology has fallen with the realization of its potentially fatal impact upon our world. We now know that nature should not have been deemed inferior because it was indescribable through Euclidean geometry. We had needed a new geometry and did not know it.

The recent discovery of fractal geometry has shown us that we cannot conclude nature is disorganized because triangles, squares, and lines cannot describe or measure it. This was an assumption based on our own limited Euclidean method. Instead of point-line-plane, fractals are based on a geometry of shapes that are "self-similar," much like the Russian nesting dolls in which a dozen dolls nest within each other, from the very large to the very small. Each individual doll, although in a different scale, is similar and therefore reflects the whole.

Unapparent to the human eye, ordered fractal shapes exist within complex natural forms. The fractal content of these structures are not obvious because the object takes on a familiar form only after a large number of iterations (repetitions) of the basic fractal "building block shapes." In order to make a fractal simulation of a mountain, depending on the level of detail, over ten million iterations may be necessary.

Although we sense them intuitively, these fractal iterations can only be objectively seen and evaluated by supercomputers. It is indeed ironic that we are dependent upon computers for this new understanding of nature.

Contrary to the thinking earlier in this century, we can now have confidence that although nature appears random, it is in fact ordered and universal. Fractal geometry allows artists to see nature in its essence of space and form in an entirely new way. This new view of nature could not have come at a more appropriate time, as we all need a new reverence for our "earth system." Fractal geometry will perhaps precipitate not only a renewed respect for nature but a progressive vision in art.

## PLANTS AS SCULPTURE?

Central to my own research and background for my sculptures has been my interest in the nearly complete absence of plant forms as subject matter in the history of sculpture.

Up until the late 20th century, plants could only be found in the decorative or architectural arts but not as sculpture. As we visualize the history of sculpture—from ancient Egypt and Greece, the Italian Renaissance, through modern art—we observe that the human or animal figures have dominated sculptural subject matter. Plants, at most, were foils for human figures, as in Bernini's *Apollo and Daphne*, done in 1623, where plants lie in subordinate position at human feet.

If it were not for decorative arts, we would hardly know today how plants appear rendered and interpreted in three dimensions. Unfortunately, the most prolific use of three-dimensional plants was not within the context of serious subject matter but in architectural decoration, e.g., capitals, cornices, ceiling plaster designs.

Michelangelo's *David*, created in 1501–1504, is the quintessential human monument. This heroic 13.5 foot figure certainly reinforces the view of human perfection. One can only speculate the leveling effect monumental plants would have had on our own anthropocentric view of the world.

Progressively, during the past 20 years, plants have been playing an increasingly important role in sculpture. My own work has been devoted to the new visual and cultural experience of plants as sculpture, both in monumental and small scale.

In the beginning, I had to create new sculpture techniques to have a capacity to make plants into bronze that would have both large scale and thinness and delicacy of surfaces. It quickly became clear to me during this experimentation that it was the physical difficulty of rendering plants that accounted for their absence in sculpture's history more than anthropocentrism.

Vasari, in his book of techniques, indirectly corroborates my conclusion, when he mentions foundries during his day (the 15th century) casting single blades of grass and small flowers as acts of virtuoso bronze casting.

I feel a distinct echo of Goethe's search for the archetypal plant—the *Urpflanze*—in my work. Goethe believed that within the structure of a single plant, all of the secrets of the universe can be discovered.

## BIBLIOGRAPHY

Barnsley, M. 1988. Fractals everywhere. Academic Press, New York.

Briggs, J. and F. D. Peat. 1989. Turbulent mirror: An illustrated guide to chaos theory and the science of wholeness. Harper & Row.

Ceysson, B. 1987. Sculpture, the great tradition of sculptures from the fifteenth to the eighteenth century. Rizzoli, New York.

Emboden, W. A. 1987. Leonardo Da Vinci on plants and gardens. Dioscorides Press.

Evan, J. 1976. Pattern, a study of ornament in Western Europe from 1180 to 1900. vol. 1. DeCapo Press.

Fleischmann, M., D. S. TiDeslay, and R. C. Ball (eds.). 1989. Fractals in the natural sciences. Princeton Univ. Press.

Henkels, H. 1987. Mondrian from figuration to abstraction. Tokyo Shimburn Museum Catalogue.

Holtzman, H. and M.S. James (eds., trans.). 1986. The new art—the new life, the collected writings of Piet Mondrian. G. K. Hall and Co., Boston.

Jaffe, H. L. C. n.d. Mondrian. Harry N. Abrams Inc., New York.

Le Normand-Romain, A. 1986. Sculpture, the adventure of modern sculpture in the nineteenth and twentieth century. Rizzoli, New York.

MacCurdy, E. (ed., trans.). 1939. The notebooks of Leonardo Da Vinci. George Braziller.

Ouspensky, P. D. 1912. A new model of the universe. Reprinted in 1971 by Vintage Books, New York.

Pedretti, C. 1980. Leonardo Da Vinci, nature studies from the Royal Library at Windsor Castle. Johnson Reprint Corporation.

Prusinkiewicz, P. 1989. Lecture notes in biomathematics, Lindenmayer systems, fractals, and plants. Springer-Verlag, New York.

CHAPTER 41

# *People and Plants:*
# *A Case Study in the Hotel Industry*

Michael R. Evans

Associate Professor of Hotel, Restaurant, and Institutional Management,
Virginia Polytechnic Institute and State University


Hollis Malone

Manger of Horticulture, Opryland Hotel, Nashville, Tennessee

## *INTRODUCTION*

The hotel and restaurant industry is one of the largest industries in the United States. Estimated 1989 gross sales in the food-service industry was $227 billion, and estimated gross sales for the hotel industry was $35 billion. The combined sales for both industries account for 5% of the U.S. Gross National Product (NRA, 1988). Each year, independent and chain-operated hotels and restaurants spend millions of dollars to design and maintain the interiors of dining rooms, lobbies, and other public space. In most cases, the impact of various design elements (e.g., lighting, furniture, fixtures, etc.) on an operation's success is virtually never evaluated. Very little data are actually collected and analyzed to determine whether various design elements have influenced the patrons' behavior and the initial selection of the property.

Flowers and plants are now very common design elements displayed in interiors of many of the leading hotel and restaurant operations in the United States. For example, the restaurant industry term "fern bar" is still used to describe the trendy dining concept developed in the 1970s, featuring a dining atmosphere filled with plants, brass, and stained glass. The heavy use of plants in the "fern bar" concept helped launch several casual chain theme restaurant concepts, including the chains T.G.I. Fridays, Bennigan's, and Houlihan's. In the lodging industry, the Hyatt hotel chain, known for its stunning atrium lobbies, has also used extensive hanging planters and baskets in a majority of its large open lobbies.

Combined with other traditional design elements of furniture, floor/wall coverings, lighting, and music, plants can be used to provide a comfortable environment that helps shape a guest's mood as well as the overall image of a facility. It is generally acknowledged that well-landscaped and maintained hospitality interiors can make guests feel more relaxed—feel a sense of sanctuary, privacy, or even luxury. Even though there is ample evidence that plants have been used extensively in many successful hospitality facilities, limited information is available related to the cost associated with purchasing and maintaining plants in hotels and restaurants and the possible benefits (at least in economic terms) from this form of investment.

Most hotel managers and designers might conclude that plants are "important product attributes" to hotel guests, but are not "determinant product attributes" (Lewis and Chambers, 1989). Important attributes include items people expect and enjoy in a hotel, including items such as bathroom amenities and/or air-conditioning. Determinant attributes are those that actually determine hotel selection or choice. These are the attributes most closely related to consumer preferences or actual purchase decisions. Determinant product attributes can be determined from several differentiating factors such as room size or quality of furnishings.

## THE OPRYLAND HOTEL: A CASE STUDY OF PLANTS AND PEOPLE

The best example of plants as a differentiating product design element and a possible determinant product attribute in the hotel industry may be the Opryland Hotel in Nashville, Tennessee. This hotel is considered one of the most financially successful meeting and convention hotels in the country. The hotel opened in 1977 with 600 guest rooms, expanded to 1068 in 1983, and expanded again to 1879 guest rooms in 1988. The hotel also has one of the largest square footages of meeting space in the country. The total investment in the hotel is approximately $145 million. As the 12th largest hotel in the country, the Opryland Hotel has received numerous honors. It has earned the Golden Key Award from *Meetings and Conventions* magazine and has been named one of the one of the ten best hotels in the country by the readers of *Corporate Meetings & Incentives* magazine. It holds both the Mobil four-star award and the AAA four-diamond award. The strength of Opryland Hotel's product is indicated by the hotel's occupancy rates of over 85% each year, well above the current national average of 68% (*Trends in the Hotel Industry*, 1989).

The hotel also has one of the largest investments in indoor and outdoor gardens in the United States. There are currently 25 acres of outdoor space and 12 acres of indoor space with approximately 18,000 indoor plants (and 600 species) valued at well over $1 million. The annual horticulture budget to maintain this living investment is approximately $1.2 million. A staff of approximately 52 tends to the plants year round.

The hotel has two massive six-story semitropical indoor gardens (actually large greenhouses). The two-acre Conservatory was completed in 1983 and the 1.5-acre Cascades was completed in 1988. Both gardens have numerous footpaths or walkways that allow guests to meander through fountains, waterfalls, and many varieties of foliage. It is estimated that over 500,000 hotel guests tour the gardens each year. Some 705 guestrooms, many with balconies, overlook the gardens in an open-seating, cafe-style restaurant that gives the impression of an old European village. These facilities were designed after studying several of the large conservatories in Canada and Europe.

The Conservatory is meant to recall the lush solitude of a Victorian garden and features a 72-foot-tall sculptured fountain called the Crystal Gazebo. There are many places to sit quietly and think. The Cascades is a water-oriented space that features a 12,500-square-foot lake and a 40-foot-tall rock mountain with three waterfalls and lush gardens. It has a fantasy-like atmosphere with a "Dancing Waters" fountain that is accented by laser beams and colored lighting in the evenings.

There are several examples of how the hotels "greatscapes" have had a positive impact on the financial success of the hotel. For example, the unusually high annual room occupancy rate of 85%, numerous awards, and the continued expansion of the complex are just a few. Another example of the positive impact on the hotel is that rooms overlooking the gardens are always the first to be reserved by repeat guests. These rooms generally command a premium price of at least $30 over rooms that do not offer garden views (1990 average daily rate of $149 per room vs. $179 per room for view of garden). This translates into an additional $7 million additional room revenue each year to the hotel. Even though a precise cost/benefit figure is very difficult to estimate, it would seem that the plant investment and maintenance costs are covered by the additional room revenue generated by the gardens.

## SUMMARY

The Opryland Hotel is one of the best examples of "plantscaping" in the hospitality industry. The large investment in plants, flowers, and landscaping has made the Opryland Hotel a unique product in the convention hotel market. Plants, as a differentiating design element, have created a positive image with many hotel meeting planners and hotel guests, and have allowed this hotel to out-perform many of its national competitors. Plants and effective landscaping may well be a determinant product attribute or a major factor for people selecting the Opryland Hotel.

## LITERATURE CITED

Lewis, R. and Chambers. 1989. Marketing leadership in hospitality. Van Nostrand Reinhold, New York.
National Restaurant Association. 1988. Forecast '89. Restaurants USA 8(11).
Trends in the Hotel Industry. 1989. Pannell Kerr Forster, Houston.

CHAPTER 42 – ABSTRACTS

# Use of Horticultural Products in the Advertising of Nonhorticultural Products: Reasons and Implications

M. Ahmedullah

Department of Horticulture and Landscape Architecture, Washington State University

## ABSTRACT

The advertisers of nonhorticultural products, like telephone and computer manufacturers, have used horticultural products in their advertisements. These horticultural products include fruits and vegetables.

This presentation deals with a survey of advertising agencies to find out the reasons for the popularity of horticultural products in advertising nonhorticultural products. Although the project is not yet complete, some of the reasons for this popularity that have been identified are (a) horticultural products like fruits are easily recognizable by the general public; (b) they attract attention easily; (c) advertisers feel that the trends in advertising are changing.

The project draws the attention of horticulturists to this phenomenon and analyzes the advertisers' reasoning in using fruits and vegetables to sell products.

# Interdisciplinary Educational Efforts
# in Horticulture

Kevin L. Grueber

Assistant Professor of Horticulture, Virginia Polytechnic Institute and State University

E. Scott Geller

Professor of Psychology, Virginia Polytechnic Institute and State University

## ABSTRACT

Students interested in pursuing careers in horticulture should be made aware of the perceptions and desires of the consumer of horticultural products. Both instructor and student may be ignorant, however, of proper procedures and protocol to determine consumer needs. In this educational project, instructors of Greenhouse Management and instructors of Human Psychology have planned a cooperative, laboratory experience for their students that will expose the students to the principles of each course. Horticulture students will grow and supply flowering crops of variable size and quality; psychology students will develop a survey questionnaire; and students will work cooperatively to gather and interpret information. Project description, procedure, and evaluations will be discussed, as well as the impact that interdisciplinary laboratory projects might have on undergraduate students.

# Potential of Interactive Video Systems as a Research Tool

Carol Ness,* Leslye Bloom,* Mary Miller,* and Diane Relf †

*Extension Interactive Design, † Associate Professor of Horticulture,
Virginia Polytechnic Institute and State University

## ABSTRACT

Public Information Interactive Systems are being used in Virginia to reach Extension clientele. Interactive video systems combine video, slides, graphics, audio, and text to provide a multimodal, user-driven information delivery system. Kiosks have been placed in 12 locations throughout Virginia, including selected malls, libraries, and one community college, with Public Information Systems developed by the Extension Design and Development Group at Virginia Tech.

Interactive video systems offer a unique opportunity to collect research data from a geographically diverse audience in conjunction with its educational purpose. We currently keep statistics allowing us to determine which Extension information is most sought after by the public. The computer can automatically keep track of the number of touches to the screen or record answers to questions in the program.

Horticulture information available on the public information systems includes a "Houseplants" program with information on 131 cut flowers and houseplants. A program on selecting landscape plants contains information on 141 trees, shrubs, vines, and ground covers. By simply touching the screen, the user may browse through photographs of plants, move through fact sheets, use the landscape plant sorter, and request print-outs on all plants in the program. The horticulture program has become the most popular of the eight information areas.

Applications to research on human issues in horticulture could include collecting data on user preferences in various areas, including plant material, landscape design features, cut flowers, sounds associated with the garden, etc. Statistics are easy to collect by counting touches made to the screen. Surveys or questionnaires including demographic information could be included with appropriate slides and graphics to illustrate the point in question.

# Interdisciplinary Research Team in Consumer Horticulture: A Research Approach

Diane Relf

Associate Professor of Horticulture,
Virginia Polytechnic Institute and State University

R. Peter Madsen

Information Officer, Office of Consumer Horticulture,
Virginia Polytechnic Institute and State University

## ABSTRACT

Interdisciplinary research is imperative for conducting research on human issues in horticulture. The development of the Interdisciplinary Research Team in Consumer Horticulture (IRTCH) has proven to be very effective here at Virginia Tech. The members of this group have sponsored a workshop on campus with approximately 100 participants and have been instrumental in conducting this symposium. The Team was established with the support of the Director of the Agricultural Research Station and the Dean of Research, and currently has 19 members from across the campus. Research projects are being conducted within several disciplines that will develop data on the impact of plants on the quality of human life.

# A LOOK AT THE FUTURE: DEVELOPING A RESEARCH INITIATIVE

CHAPTER 43

# Summary of the Agenda-setting Meeting

## INTERNAL AND EXTERNAL SUPPORT/OBSTACLES AND FEASANCE TO IMPLEMENT A HUMAN ISSUES IN HORTICULTURE RESEARCH INITIATIVE

More than 30 prominent members of the horticultural community and allied fields met on 22 April 1990 to discuss the future needs for research in Human Issues in Horticulture (HIH). Led by a facilitator, participants first reached a consensus that the time had arrived for cooperating to address these issues. During the morning session, they analyzed the following elements of organizing a research effort:

- the external (non-horticulture) support and obstacles to establishing a commonly held research goal (Table 1);

- the internal (horticulture community) support and obstacles to establishing a commonly held research goal (Table 2);

- the current status regarding feasance, or ability to organize (Table 3).

## ACTION PLANS: TWO-FOLD APPROACH TO STRUCTURING/ORGANIZING FOR ACTION

### Work Within the Existing Associations

Encourage all horticultural community (HC) associations to support Human Issues in Horticulture (HIH) activities within the context of their existing activities. Publish articles in newsletters, solicit research articles for journals, conduct workshops/presentations at conferences, and support research through endowments.

Hold meetings including officers, staff, and leaders at annual conferences of the existing associations to communicate the importance of HIH and encourage research and/or funding of research. (Note: The American Society for Horticultural Science, the American Association of Botanical Gardens and Arboreta, and the American Horticultural Therapy Association have special workshops or presentations addressing HIH at their 1990 annual conferences.)

**Table 1.** External (non-horticulture) factors influencing the establishment of a research initiative.

| Support/factors that increase the potential of success | Obstacles/factors that reduce the potential of success |
|---|---|
| Awareness<br>　environment/nature<br>　horticulture/gardening<br>　positive perception | Awareness limited<br>　HIH is unrecognized<br>　perception that horticulture is work<br>　lack of hard data |
| Economic<br>　increased buying power in certain<br>　　population segments<br>　retirement | Economic<br>　competition for discretionary money<br>　not enough known about cost/benefit |
| Quality of Life (QOL)<br>　pursuit of pleasure and tranquility<br>　aging society/older values | Quality of Life (QOL)<br>　competition with arts/cultural |
| Groups<br>　communications/media<br>　tourism<br>　leisure<br>　human services<br>　environmental<br>　development (design/build) | Groups<br>　media—lack of information<br>　developers—could increase their costs |
| Motivation<br>　profit<br>　QOL<br>　environmental health | Motivation<br>　cost/benefit<br>　competition<br>　lack of hard data |
| Timing is right<br>　1990s: the decade of the<br>　　environment<br>Interdisciplinary nature of HIH<br>Social sciences<br>Marketing, tourism, recreation<br>Medical profession<br>Some existing methodology | Water restrictions<br>Limited government budgets<br>Vested interests<br>Cost/benefit<br>Lack of action<br>Lack of hard data<br>Lack of universal support<br>Lack of knowledge regarding quality<br>　and professionalism<br>　of human-based research in horticulture<br>Territorialism (social science vs. plant science)<br>Cost/benefit<br>Lack of hard data |

**Table 2.** Internal (horticultural) factors influencing the establishment of a research initiative.

| Support/factors that increase the potential of success | Support/factors that reduce the potential of success |
|---|---|
| Commitment | Lack of precedent |
| Diversity/richness of backgrounds | Commodity orientation of horticulture community |
| Interdisciplinary nature | Lack of identified, fundable projects |
| Have some hard data | Lack of respect for methodology and |
| Recognized need for HIH data | findings of other disciplines |
| Common goal (HIH) | Scope too broad |
| Industry has started funding research | Territorialism is barrier to interdisciplinary research |
| Intuitive understanding of people-plant interaction | Failure to understand need for research |
| Other research has commenced | Not a high priority |
| Presents opportunity for expansion | Wide territorialism; failure to focus on |
| Have increased visibility | specific issues of some specialty |
| Have energy/enthusiasm to communicate | groups (AABGA, AHTA, ACGA, etc.) |
| Cost/benefit ratio good | Internal competition |
| | Perceived incompatibility in our diversity |
| | Lack of image of horticultural professional |
| | Product/plant orientation |
| | Horticultural curricula too narrow/lack |
| | of interdisciplinary courses |
| | Failure to understand need for research |
| | Confusion with Hort Therapy |
| | Nontraditional nature |
| | Lack of unique identity for HIH |
| | Lack of interpretation |
| | Lack of dissemination |
| | Lack of interdisciplinary communication |
| | Hard to lobby |
| | Hard to mainstream |
| | Lack of more cost/benefit data |
| | Lack of jobs/students |
| | Lack of recognition of the need for HIH |
| | trained professionals |

Strength summary: dedication, intuition, diversity with common goals, positive cost/benefit, high visibility field.

**Table 3.** Do we possess the feasance to organize?

| | Do we have the: | | |
|---|---|---|---|
| Resources | Methodology | Authority | Other strengths |
| funding—not now, but possible | Coordination consortium/ council—not a | administrative support—not now, but possible | shared goals |
| people—yes, a start | | | positive political and public |
| time—possibly; if funding there | new organization network exists; funding needed to continue | leadership—can get | support and media interest |
| | Communications commitment | | |

**Examples of Possible Activities**

The American Society for Horticultural Science has two publications, *HortTechnology* and *HortScience*, that could appropriately have Associate Editors for Human Issues in Horticulture. Horticulture Research Institute and American Floral Endowment could be encouraged to expand their funding to HIH research. The American Association of Nurserymen could be encouraged to have an Associate Editor for HIH in the *Journal of Environmental Horticulture*. Trade and professional associations can include HIH in inviting speakers for conferences and choosing articles for their publications.

## Establish a Human Issues in Horticulture Consortium or Council

Organize or structure an HIH consortium/council for communication and research.

**Mechanisms and Considerations for Organizing**

Establish an ad hoc advisory group:

A. Members—to be representatives of consortium sponsors and local (Washington, D.C.) groups from the horticultural community.

B. Purpose—to address issues defined by the 22 April 1990 agenda-setting group:
   1. define "world of human horticulture" categories (identify related associations, key players)
   2. solicit official representatives from the entire horticultural community
   3. establish a steering committee/core group
   4. establish goals/mission
   5. establish a formalized procedure
   6. develop a name for this entity
   7. decide on a home for this entity
   8. identify resources for staff and communication expenses; make financial commitments.

C. Ad hoc meeting to be held 24 May 1990 at AAN offices in Washington, D.C. and attended by Jim Swasey, American Association of Botanical Gardens and Arboreta; Chuck Richman, American Horticultural Therapy Association; Larry Scovotto, American Association of Nurserymen; Skip McAfee, American Society for Horticultural Science; Marvella Crabb, Society of American Florists; Earl Wells, Florida Nurserymen and Growers Association; Diane Relf, Candice Shoemaker, and Pete Madsen, Virginia Polytechnic Institute and State University.

## Establishing Communication

A. Circulate press releases to the horticulture community (HC) and the public media—initial mailing 30 April 1990.

B. Send a summary of the symposium *The Role of Horticulture in Human Well-Being and Social Development* (4 to 8 pages) to:
   1. sponsors/funding sources/endorsers
   2. those who request information
   3. those from whom we solicit funds
   4. interested associations, to encourage participation.

C. Identify specific publications to carry HIH articles and set time lines for appearance of articles (within six months):
   1. *HortScience* (Virginia Lohr, Lynn Doxon)
   2. *Journal of Horticultural Therapy*

3. trade publications
   *American Nurseryman* (Diane Relf)
4. other: non-horticulture.

D. Proceedings to Timber Press by 30 June 1990 for publication in 1991.

E. Bring the agenda-setting group from 22 April 1990 back together in six months (Roy Taylor volunteered the Chicago Botanic Garden).

F. Organize a second symposium on "Challenges and Opportunities in HIH" with an emphasis on appropriate methodologies and technology—to be held at the Chicago Botanic Garden or Ameriflora '92 in Columbus? Jules Janick was suggested to develop the program.

## Establishing a System to Increase Research

A. Conduct a "think tank" of HIH researchers to develop a list of feasible research projects of high quality and impact (i.e., benefits of plants in exterior and interior areas of homes for the aged). Consider topics for which research would lead to documentation of improvements in the quality of human life.

B. Establish a database/network to facilitate sharing of current knowledge.

C. Establish a communication network to relay research information to users as rapidly as possible.

D. Establish a research system within the consortium through use of two subgroups to identify funds and to solicit/evaluate research proposals.

E. Work with the USDA to include HIH in their research budget.

F. Maintain legislative contacts and education to begin development of legislative action regarding HIH research (as per Charles Hess' comments at the welcome address on 19 April 1990).

CHAPTER 44

# Development of the People-Plant Council

---

## 24 MAY 1990 MEETING OF THE AD HOC COMMITTEE ON HUMAN ISSUES IN HORTICULTURE

The designated committee met on 24 May 1990. After briefly sharing reports of the many follow-up activities resulting from the symposium, the committee addressed the issues set forth at the 22 April 1990 meeting. A unanimous agreement was reached as to the need for a steering entity, and the People-Plant Council was formed. The Council's mission and strategy for achieving that mission were set forth, and the composition of, and participation in, the Council was discussed.

## 7 OCTOBER 1990 MEETING OF THE ORGANIZING COMMITTEE OF THE PEOPLE-PLANT COUNCIL

Representatives of various groups interest in PPC met at the Chicago Botanic Gardens. Efforts and accomplishments of the individual associations were discussed. The prospectus was refined and the final report for the symposium was presented. The representatives from the Society of American Florists agreed to outline a strategic plan for consideration at the next meeting. The name of the PPC was changed to People-Plant Counsel to emphasize that its mission is providing information, not serving as simply another association.

## 15 MARCH 1991 MEETING OF THE ORGANIZING COMMITTEE OF THE PEOPLE-PLANT COUNCIL

Held in Alexandria, Virginia at the offices of the Society of American Florists, the meeting was well attended with 14 associations or agencies represented.

The mission statement and strategic plan were further clarified and adopted. Copies of these documents follow. The name of this group was agreed upon as People-Plant Council, to avoid confusion from the unusual use of the word Counsel. Questions of management of the PPC were clarified by electing Diane Relf as the Coordinator. A proposed budget and explanation of the role of Virginia Polytechnic Institute and State University were requested. The first newsletter was reviewed and approved to be mailed. Priorities for immediate action were established. It was agreed that information targeting the specific benefits to individual associations in affiliating with PPC would be developed and an individual identified to present the information to the boards of each of the associations considered for membership. It was agreed that the PPC steering/planning committee should continue to meet twice a year, in spring and fall. Steve Daigler of the Society of American Florists agreed to coordinate the schedules of the various associations to determine an appropriate time for our next two meetings.

CHAPTER 45

# *The People-Plant Council Mission, Strategy, and Affiliation*

---

The People-Plant Council (PPC) was formed on 24 May 1990 as a direct result of the national interdisciplinary symposium, "The Role of Horticulture in Human Well-Being and Social Development," held in April 1990.

## *MISSION*

The mission of this Council is to document and communicate the effect that plants and flowers have on human well-being and quality of life. This mission is to be carried out through the following five-part strategy focusing on the effects that plants have on human well-being through the psychological, sociological, physiological, economic, and environmental effects they produce:

1. Communication—maintain an interdisciplinary network among researchers, funding sources, users, and Council affiliates; provide research-based information to the horticulture and social science communities, commercial and private users of plants, and the general public.

2. Research—encourage cooperative efforts in identifying research priorities and establishing interdisciplinary research methodologies.

3. Funding—establishing a network to link researchers to funding sources, including government agencies, public and private foundations, and co-operatives.

4. Implementation—encourage the use of horticulture for enhancing the quality of life based on research findings and provide consulting services to users to implement research data.

5. Education—encourage the development of a public curriculum to include people-plant interaction as an essential subject in kindergarten through adult/continuing education in many fields of study.

The PPC serves as a link between organizations representing all facets of the horticulture and the social science communities. The Council is a network designed by the representatives of interested associations to enhance and focus their efforts toward documenting the human benefits derived from horticulture.

## AFFILIATION

Affiliation with the PPC is established through contributions to maintain its operational expenses. Contributions are based on the size and scope of the affiliating organization. All contributions are handled through Virginia Polytechnic Institute and State University through its Industrial Affiliates Program in the office of Grants and Contracts, and therefore are tax deductible.

Affiliation is open to all organizations within the horticulture and social science community and allied or interested organizations to include, but not be limited to:

- academic and professional associations of many disciplines;

- trade and commercial associations;

- volunteer, civic, amateur, and concerned groups.

Affiliation entitles the organization or association to:

- send an official representative to all People-Plant Council meetings to participate in agenda setting, identification of targeted research initiatives, selection of an advisory committee, determination of communication and implementation focus, and other business of the Council;

- use the name and logo of the People-Plant Council in their organization's official publications to identify themselves as a supporter of the Council's mission of enhancing the quality of human life through horticulture;

- designate an individual to receive all communications of the Council and become an active member of the Council's research-communications network, thus having primary access to people-plant interaction research data for the development of new research, the application of research to enhance quality of life, and the promotion of their segment of the horticulture community.

Affiliates will be acknowledged on the letterhead, annual reports, and other appropriate publications of the People-Plant Council.

## CONTRIBUTORS

In addition to organizational affiliates, the Council seeks funding from contributors. Contributors include commercial horticulture businesses, public relations and consulting firms, foundations, endowments, and individuals who have a commitment to the vision of the Council and seek to support its goals and fund its operational strategies.

Contributors are entitled to:

- send a representative to conferences, training sessions, and other educational programs of the People-Plant Council at a discounted rate;

- use the name and logo of the People-Plant Council in their organization's official publications to identify themselves as a supporter of the Council's mission of enhancing the quality of human life through horticulture.

- designate an individual to receive all communications of the Council and become an active member of the Council's research-communications network, thus having primary access to PPI research data for the development of new research, the application of research to enhance quality of life, and the promotion of the contributors' sphere of the horticulture community.

## USERS

Researchers, educators, and others can use the services of the People-Plant Council, which will include a biannual newsletter, periodic update reports, access to a computerized information network/database, and future conference/educational program registration with a cost-of-service fee.

CHAPTER 46

# Strategic Plan
# of the People-Plant Council:
# Linking Horticulture
# with Human Well-Being

---

## MISSION

To document and communicate the effect that plants and flowers have on human well-being and life-quality.

## OBJECTIVES

### Communication

To communicate on an ongoing basis the psychological, sociological, physiological, economic, and environmental effects of plants and flowers to the academic community to include horticulture, social sciences, and other interested disciplines, the commercial and private users of plants, and the general public.

**Strategies**

- Originate editorial themes for stories, articles, speeches, videos, etc.
  Maintain information on status of research and findings (both pre- and post-publication). Provide access to copies of the publication when possible.

- Identify and provide vehicles for dissemination of research information to the above target audiences, accessing and working cooperatively with existing groups.
  Establish methods of distribution of information to include press releases, newsletters, electronic networks, presentations at conferences, and magazine articles to user/consumer publications.

- Produce and distribute communications.

## Research

To provide direction for people-plant research.

**Strategies**

- Develop comprehensive lists of existing research.

- Identify research priorities.
  Conduct surveys of users of horticultural crops to determine the research information they need to enhance their uses and optimize the impact on the quality of human life.
  Hold "think tank" seminars with invited participants representing various professions that have an intrinsic interest in people-plant interaction.

- Communicate research priorities to researchers through their professional associations (i.e., American Society for Horticultural Science, American Horticultural Therapy Association, American Association of Botanical Gardens and Arboreta).
  Identify working groups or committees addressing human issues in horticulture in professional associations.
  Communicate research priorities at conferences, in journals and newsletters, and through other established communication networks.

- Provide support toward the establishment of new research projects.
  Provide and/or encourage workshops on interdisciplinary research and appropriate research methodologies that bring together researchers from diverse settings.
  Assist in proposal development and linkage to funding sources.

## Funding

To increase funding for people-plant research and dissemination of the research results.

**Strategies**

- Identify potential sources of research funds both inside and outside the horticulture community.

- Work with research funding sources to increase money designated for people-plant research.
  Educate funding groups as to the value of this type of research and cultivate interest in funding people-plant research.
  Provide linkages to researchers.
  Keep funding groups informed of the broad impact of people-plant research and its application.

- Establish an information clearinghouse on funding sources.

- Publicize the availability of funding information via industry and professional association communications.

## Implementation

To increase the use of horticulture to enhance quality of life based on research findings.

**Strategies**

- Provide consulting services to users of horticultural crops in order to implement research findings.

    Identify agencies, companies, and other organizations that could effectively utilize horticulture to enhance quality of life (i.e., health agencies, tourist industries, urban planners and developers).

    Educate these users or potential users of the benefit to them and their clients of the uses of plants.

    Offer consulting services to identify appropriate research data and assist in incorporating the information into their operational plans.

- Work with staff of affiliated associations to encourage the associations to take action to support the mission of PPC.

    Identify key personnel and association goals compatible with PPC goals.

    Identify specific tasks that require association action for implementation.

    Provide staff with information and strategies appropriate to support action requested.

- Provide assistance for publication in this field.

    Identify influential publications (i.e., trade publications for developers, theme park managers).

    Educate editors as to the value of publishing related information.

    Inform researchers, Extension personnel, and garden writers of the potential publishing market.

## *Education*

Encourage the incorporation of people-plant information into the educational materials, programs, and curricula at all academic levels.

**Strategies**

- Work with educational institutions for acceptance of information on people-plant interaction in the curricula of horticulture and other disciplines.

    Educate faculty at universities and colleges as to the importance of including this information in their curricula.

    Act as a resource on this information to be used in introductory horticulture or other courses.

- Identify producers of educational materials.

- Provide significant data and teaching aids.

## *Establish Organizational Competency*

To maintain an organizational structure for the People-Plant Council that efficiently uses the available resources and personnel to address its mission effectively.

- Maintain linkages with all affiliates.

- Conduct planning sessions and "think tanks" with affiliates or other designated professionals to maintain goals and direction.

- Manage staff, budget, and fund raising for Council operations.

# SECTION VIII

# *APPENDIXES*

APPENDIX A

# *Symposium Overview*

---

## The Role of Horticulture in Human Well-Being and Social Development: A National Symposium

### *19–21 April 1990*

HONORARY SYMPOSIUM CHAIRS:
>    Sen. Mark Hatfield (Oregon)
>    Rep. Jamie Whitten (Mississippi)

SPONSORS:
>    Department of Horticulture, Virginia Polytechnic Institute and State
>        University
>    American Society for Horticultural Science
>    American Association of Botanical Gardens and Arboreta
>    American Horticultural Therapy Association

SYMPOSIUM CHAIR:
>    Diane Relf, Virginia Polytechnic Institute and State University

SYMPOSIUM ORGANIZING COMMITTEE:
>    Charles Lewis, Morton Arboretum
>    Harold B. Tukey, Jr., University of Washington
>    Skip McAfee, American Society for Horticultural Science
>    Susan Lathrop, American Association of Botanical Gardens and Arboreta
>    Chuck Richman, American Horticultural Therapy Association

CONFERENCE COORDINATOR:
>    R. Peter Madsen, Office of Consumer Horticulture, Virginia Polytechnic
>        Institute and State University

CONFERENCE MODERATOR:

> Thomas Fretz, Associate Dean and Director, Iowa Agriculture and Home Economics Experiment Station, Iowa State University, President-elect, American Society for Horticultural Science

FUNDED IN PART BY:

> American Floral Endowment
> Associated Landscape Contractors of America
> Bailey Nurseries
> Ciba-Geigy Corporation
> Conard-Pyle Company
> Costa Nursery Farms
> Dramm International
> Florida Nurserymen and Growers Association
> Garden Writers Association of America
> Horticultural Research Institute
> Jackson & Perkins Co.
> Kenneth Post Foundation
> Noma Outdoor Products
> O. M. Scott & Sons
> Paul Ecke Poinsettias
> Rain Bird
> Rhone-Poulenc
> U.S. Department of Agriculture-Cooperative State Research Service

ENDORSED BY:

> American Association of Nurserymen
> American Community Gardening Association
> American Society of Consulting Arborists
> American Sod Producers Association
> American Forestry Association
> Associated Landscape Contractors of America
> Horticultural Research Institute
> National Gardening Association
> National Greenhouse Manufacturers Association
> National Urban Forest Council
> North American Horticultural Supply Association
> Nursery Industry Association of Australia
> Professional Grounds Management Society
> Professional Plant Growers Association
> Society of American Florists
> Virginia Greenhouse Growers Association
> Virginia Nurserymens' Association
> U.S. Botanic Garden
> Wholesale Florists and Florist Suppliers of America

# APPENDIX B

# *Moderators' and Authors' Addresses*

---

## MODERATORS

### Conference

Thomas Fretz
Associate Dean and Director
Iowa Agr. and Home Economics
    Experimental Station
Iowa State University
Ames, IA 50011

### Sessions

Barry Adler
O. M. Scott & Sons
Landscape Services
Marysville, OH 43041

David Bradshaw
Department of Horticulture
Clemson University
Clemson, SC 29631

Joel S. Flagler
Bergen City Agriculture/Resource
    Management
327 Ridgewood Avenue
Paramus, NJ 07652-4896

Virginia I. Lohr
Horticulture and Landscape Architecture
Washington State University
Pullman, WA 99164-6414

Candice A. Shoemaker
Department of Agriculture
Berry College
5003 Mount Berry Station
Rome, GA 30149-5003

Doug Welsh
225 Horticulture/Forest Science Bldg.
Texas A&M University
College Station, TX 77843

## AUTHORS

M. Ahmedullah
Horticulture and Landscape Architecture
Washington State University
Pullman, WA 99164-6414

Douglas L. Airhart
Dept of Agriculture
Tennessee Technological University
Cookeville, TN 38505

James A. Azar
Veterans Administration Medical Center
    ɔrthampton, MA 01060-1288

John C. Billing
Texas Technical University
Park Administration and Landscape
    Architecture
P.O. Box 4169
Lubbock, TX 79409

Leslye Bloom
Extension Interactive Design
Virginia Polytechnic Institute and State
    University
Blacksburg, VA 24061

Blaine Bonham, Director
Philadelphia Green
325 Walnut St
Philadelphia, PA 19106-2777

Sue Burd Brogden
Director of Education
Chicago Botanic Garden
P.O. Box 400
Glencoe, IL 60022-0400

Charlene A. Browne
Landscape Architecture
Virginia Polytechnic Institute and State
    University
Blackburg, VA 24061

A. Jerry Bruce
Sam Houston State University
Division of Psychology and Philosophy
Huntsville, TX 77341

Magne Bruun
College of Landscape Architecture
Agriculture University of Norway
Norway

Clifton Bryant
Department of Sociology
Virginia Polytechnic Institute and State
    University
Blacksburg, VA 24061

Bruce Butterfield
National Gardening Association
180 Flynn Avenue
Burlington, VT 05401

David J. Chalmers
Department Horticultural Science
Massey University
Palmerston North
New Zealand

Nancy K. Chambers
Glass Garden, Rusk Institute of
    Rehabilitation Medicine
400 East 34th St.
New York, NY 10016

Thomas Conroy
Communications Department
University of Massachusetts
Amherst, MA 01003

Chris Cordts, Coordinator
Denver Urban Gardening Program
Colorado State University, Denver
    Cooperative Extension
1700 S. Holly St.
Denver, CO 80222

Joseph C. Cremone, Jr.
Harvard University
160 Commonwealth Ave.
Boston, MA 02116

Richard P. Doherty
52 Chestnut St.
Amherst, MA 01002

Lynn Ellen Doxon
Cooperative Extension Service
U.S. Department of Agriculture
New Mexico State University
Las Cruces, NM 88003-0031

Kathleen M. Doutt
Bryn Mawr Rehabilitation Hospital
414 Pavili Pike
Malvern, PA 19355

Robert G. Dyck
Professor of Urban Affairs and Planning
Virginia Polytechnic Institute and State
    University
Blacksburg, VA 24061

Catherine Eberbach
New York Botanical Garden
Bronx, NY 10458

Michael R. Evans
Department of Hotel, Restaurant, and
    Institutional Management
Virginia Polytechnic Institute and State
    University
Blacksburg, VA 24061

Joel S. Flagler
Bergen City Agriculture/Resource
    Management
327 Ridgewood Avenue
Paramus, NJ 07652-4896

Mark Francis
Environmental Design
University of California
Davis, CA 95616

Bilge Friedlaender
502 Westview
Philadelphia, PA 19119

E. Scott Geller
Department of Psychology
Virginia Polytechnic Institute and State
    University
Blacksburg, VA 24061

Kevin L. Grueber
Department of Horticulture
Virginia Polytechnic Institute and State
    University
Blacksburg, VA 24061-0327

James E. Healy
Department of Psychology
Virginia Polytechnic Institute and State
    University
Blacksburg, VA 24061

Deborah B. Hill
Dept of Forestry
University of Kentucky
Lexington, KY 40546-0073

Mary K. Honeyman
Botanica
701 Amidon
Wichita, KS 67203

Sally Hoover
Horticultural Therapy
Chicago Botanic Garden
P.O. Box 400
Glencoe, IL 60022-0400

R. Bruce Hull, IV
College of Architecture
Texas A&M University
College Station, TX 77843

Jules Janick
Department of Horticulture
Purdue University
West Lafayette, IN 47907

Rachel and Stephen Kaplan
School of Natural Resources
Dena Building
University of Michigan
Ann Arbor, MI 48109-1115

Charles A. Lewis
Morton Arboretum
Lisle, IL 60532

Virginia I. Lohr
Horticulture and Landscape Architecture
Washington State University
Pullman, WA 99164-6414

Marion B. MacKay
Department of Horticultural Science
Massey University
Palmerston North
New Zealand

R. Peter Madsen
Department of Horticulture
402 Saunders Hall
Virginia Polytechnic Institute and State
   University
Blacksburg, VA 24061-0327

Hollis Malone
Manager of Horticulture
Opryland Hotel
Nashville, TN

Eisuke Matsuo
Faculty of Agriculture
Kagoshima University
Kagoshima-shi 890
Japan

Richard Mattson
Department of Horticulture
Waters Hall
Kansas State University
Manhattan, KS 66506

Brian G. McDonald
Sam Houston State University
Division of Psychology and Philosophy
Huntsville, TX 77341

Mary G. Miller
Extension Interactive Design
Virginia Polytechnic Institute and State
   University
Blacksburg, VA 24061

Anupam Mukherjee
Urban Affairs and Planning
Virginia Polytechnic Institute and State
   University
Blacksburg, VA 24061

Konrad R. Neuberger
Koelnerstr. 82-49
4018 Langenfield
West Germany

Carol Ness
Extension Interactive Design
Skelton House
207 Roanoke St.
Virginia Polytechnic Institute and State
   University
Blacksburg, VA 24061

Lucy Nguyen-Hong-Nhiem
Bilingual Collegiate Program
Univ of Massachusetts
Amherst, MA 01003

Dana C. Parker
2833 River Rd
Virginia Beach, VA 23454

Russ Parsons
College of Architecture
Texas A&M University
College Station, TX 77843-3137

Ishwarbhai C. Patel
Rutgers Extension Urban Gardening
249 University Ave., SSB Room 201
Newark, NJ 07102

Kim Randall
Department of Psychology
Virginia Polytechnic Institute and State
   University
Blacksburg, VA 24061

Diane Relf
Department of Horticulture
Virginia Polytechnic Institute and State
   University
Blacksburg, VA 24061-0327

James W. & Catherine M. Reuter
Program and Family Support
Bancroft School
Hopkins Lane
Haddonfield, NJ 08033

Jay Stone Rice
1330 Lincoln, Suite 309
San Rafael, CA 94901

Lawrence C. Rosenfield
Communication Arts, Queens College
City University of New York
535 East 80th St.
New York, NY 10021

Reginald Shareef
SCORE! Management Consultants
P.O. Box 5161
Roanoke, VA 24012

Rhonda Roland Shearer
74 Fifth Avenue
New York, NY 10011

Candice A. Shoemaker
Department of Agriculture
Berry College
5003 Mount Berry Station
Rome, GA 30149-5003

John Tristan
Department of Plant and Soil Sciences
University of Massachusetts
Amherst, MA 01003

Roger S. Ulrich
College of Architecture
Texas A&M University
College Station, TX 77843-3137

Gabriela Vigo
College of Architecture
Texas A&M University
College Station, TX 77843-3137

Patrick Neal Williams
Glass Garden, Rusk Institute of
    Rehabilitation Medicine
400 E 34th St.
New York, NY 10016

Sara Williams
Department of Horticulture Science
Kirk Hall, University of Saskatchewan
Saskatoon, Saskatchewan S7N 0W0
CANADA

# *Index*

This index, composed specifically for the Human Issues in Horticulture researcher, is a collection of key words and concepts relating to the interdisciplinary field of people-plant interaction.